Proceedings

Ein stetig steigender Fundus an Informationen ist heute notwendig, um die immer komplexer werdende Technik heutiger Kraftfahrzeuge zu verstehen. Funktionen, Arbeitsweise, Komponenten und Systeme entwickeln sich rasant. In immer schnelleren Zyklen verbreitet sich aktuelles Wissen gerade aus Konferenzen, Tagungen und Symposien in die Fachwelt. Den raschen Zugriff auf diese Informationen bietet diese Reihe Proceedings, die sich zur Aufgabe gestellt hat, das zum Verständnis topaktueller Technik rund um das Automobil erforderliche spezielle Wissen in der Systematik aus Konferenzen und Tagungen zusammen zu stellen und als Buch in Springer.com wie auch elektronisch in Springer Link und Springer Professional bereit zu stellen.
Die Reihe wendet sich an Fahrzeug- und Motoreningenieure sowie Studierende, die aktuelles Fachwissen im Zusammenhang mit Fragestellungen ihres Arbeitsfeldes suchen. Professoren und Dozenten an Universitäten und Hochschulen mit Schwerpunkt Kraftfahrzeug- und Motorentechnik finden hier die Zusammenstellung von Veranstaltungen, die sie selber nicht besuchen konnten. Gutachtern, Forschern und Entwicklungsingenieuren in der Automobil- und Zulieferindustrie sowie Dienstleistern können die Proceedings wertvolle Antworten auf topaktuelle Fragen geben.

Today, a steadily growing store of information is called for in order to understand the increasingly complex technologies used in modern automobiles. Functions, modes of operation, components and systems are rapidly evolving, while at the same time the latest expertise is disseminated directly from conferences, congresses and symposia to the professional world in ever-faster cycles. This series of proceedings offers rapid access to this information, gathering the specific knowledge needed to keep up with cutting-edge advances in automotive technologies, employing the same systematic approach used at conferences and congresses and presenting it in print (available at Springer.com) and electronic (at Springer Link and Springer Professional) formats.
The series addresses the needs of automotive engineers, motor design engineers and students looking for the latest expertise in connection with key questions in their field, while professors and instructors working in the areas of automotive and motor design engineering will also find summaries of industry events they weren't able to attend. The proceedings also offer valuable answers to the topical questions that concern assessors, researchers and developmental engineers in the automotive and supplier industry, as well as service providers.

Weitere Bände in der Reihe http://www.springer.com/series/13360

Torsten Bertram
(Hrsg.)

Fahrerassistenzsysteme 2018

Von der Assistenz zum automatisierten Fahren 4. Internationale ATZ-Fachtagung Automatisiertes Fahren

Springer Vieweg

Hrsg.
Torsten Bertram
Technische Universität Dortmund
Dortmund, Deutschland

Ergänzendes Material zu diesem Buch finden Sie auf http://extras.springer.com.

ISSN 2198-7432 ISSN 2198-7440 (electronic)
Proceedings
ISBN 978-3-658-23750-9 ISBN 978-3-658-23751-6 (eBook)
https://doi.org/10.1007/978-3-658-23751-6

Die Deutsche Nationalbibliothek verzeichnet diese Publikation in der Deutschen Nationalbibliografie; detaillierte bibliografische Daten sind im Internet über http://dnb.d-nb.de abrufbar.

Springer Vieweg

Verantwortlich im Verlag: Markus Braun

Springer Vieweg ist ein Imprint der eingetragenen Gesellschaft Springer Fachmedien Wiesbaden GmbH und ist ein Teil von Springer Nature
Die Anschrift der Gesellschaft ist: Abraham-Lincoln-Str. 46, 65189 Wiesbaden, Germany

Vorwort

Die vierte ATZ-Fachtagung „Fahrerassistenzsysteme – Von der Assistenz zum automatisierten Fahren“ 2018 zeigt die Entwicklungen hin zum automatisierten Fahren und die hierfür erforderlichen Kompetenzen in fünf Übersichtsbeiträgen, einer Podiumsdiskussion zum Thema „Automatisiertes Fahren – technische Entwicklungen im Spannungsfeld von Politik, Recht und Ethik“ und in sechs Themengebieten: Sensoren, Versicherungskonzepte, Safety und Security, Neue Methoden, Anwendungen im Nutzfahrzeug und Machine Learning.

Innovative Assistenzsysteme lassen heute Fahrzeuge bereits sehen – Wahrnehmung der Umgebung über Sensoren –, denken – vorausschauend und situationsgerecht einen angemessenen Handlungsablauf über Situationsinterpretation, Schlussfolgerung, Planung, Planerkennung, Kommunikation sowie Kollaboration zu generieren – und handeln – Handlungen sicher und zuverlässig über Aktoren ausführen –. Bei der Selbstregulation (Wahrnehmung und Interpretation, Lernen und Schlussfolgern, Planung und Planerkennung, Kommunikation und Kollaboration) kann die Künstliche Intelligenz einen bedeutungsvollen Beitrag leistet. Mit dem Machine Learning ist dem automatisiert fahrenden Fahrzeug ein Werkzeug gegeben, dass bei einer umfangreichen Anzahl an Messungen aus diesen Daten komplexe Zusammenhänge erkennt. Mit zunehmender Integration der Künstlichen Intelligenz in das automatisiert fahrende Fahrzeug wird dieses mehr und mehr ohne menschliche Steuerung sowie Regelung und detaillierte, situationsausgerichtete Programme ein vorgegebenes Ziel selbstständig und an die aktuelle Situation angepasst erreichen.

Interoperabilität ist die Grundvoraussetzung für die Einführung hoch automatisierten Fahrens. Zahlreiche Diskussionen drehen sich aktuell um die Abgrenzung von Level 3 und 4 und vor allem um die Frage, welche fahrfremden Tätigkeiten der Fahrer im automatisierten Modus wie ausüben darf. Geschehen diese auf dem Bildschirm des Kombiinstruments, so ist die Aufmerksamkeit des Fahrers zur Übernahme der Fahraufgabe durch Um- bzw. Abschalten des Bildschirms einfacher zu erreichen. Mobile Smart Devices erfordern eine andere Strategie. Weitere Aspekte sind der Fahrer (Mensch) in der Interaktion mit dem Fahrzeug (Maschine) sowie die damit verbundenen Interdependenzen.

Ein weiterer Schwerpunkt auf dem Weg zum automatisierten und fahrerlosen Fahren liegt neben den rechtlichen Rahmenbedingungen und den neuartigen Versicherungskonzepten bei der Evaluation der automatisierten Fahrfunktionen. Hierunter fallen nicht nur die Tests während der Entwicklung und Fahrzeugzulassung, sondern auch der Verifikation sowie Validierung im Rahmen der regelmäßigen technischen Überwachung sind bereits im Rahmen der Entwicklung die erforderliche Aufmerksamkeit zu schenken.

Die ATZ-Fachtagung Fahrerassistenzsysteme 2018 hat den Stand der Entwicklungen im oben aufgezeigten Themenspektrum in einzelnen Punkten dokumentiert und im Rahmen von zahlreichen Diskussionen im Kontext der Beiträge auch neue Fragen formuliert, die die Weiterentwicklung sicherlich mit beeinflussen.

Die Tagung 2018 erbrachte wieder zahlreiche wichtige Ergebnisse auf dem Weg vom assistierten zum automatisierten Fahren, die in den einzelnen Vortragsmanuskripten oder Vortragsfolien weiter ausgeführt sind. Der vorliegende Band enthält die Beiträge, die zur Veröffentlichung freigegeben worden sind.

Univ.-Prof. Dr.-Ing. Prof. h.c. Dr. h.c. Torsten Bertram
Technische Universität Dortmund
Wissenschaftliche Leitung der Tagung

Inhaltsverzeichnis

Radar for Autonomous Driving – Paradigm Shift from Mere Detection to Semantic Environment Understanding

Jürgen Dickmann(✉), Jakob Lombacher, Ole Schumann, Nicolas Scheiner, Saeid K. Dehkordi, Tilmann Giese, and Bharanidhar Duraisamy

Daimler AG, Sindelfingen, Germany
juergen.dickmann@daimler.com

1 Introduction

The challenges for sensors and their correlated perception algorithms for driverless vehicles are tremendous. They have to provide more comprehensively than ever before a model of the complete static and dynamic surroundings of the ego-vehicle to understand the correlation of both with reference to the ego-vehicle's movement. For dynamic objects, this means that radar has to provide the dimension and complete motion state as well as the class information, in highway, rural, and inner city scenarios. For the static world, new algorithm schemes have to be developed to enhance the shape representation of an object by image like semantics. In order to generate the necessary information, radar networking for 360° coverage have to be reinvented. Radar data processing toolchains have to be revolutionized by applying artificial intelligence and advanced signal processing in a synergetic manner.

The paper will provide an overview of state-of-the-art automotive radar usage, deduces future requirements for highly automated driving and will present recent advances in radar based environmental perception.

1.1 The History of Automotive Radar Answers the Question: "Why Radar?"

Automotive radar has already reached a market penetration that leads to several tens of million units used. It has grown up to a status where it has found its way into nearly all car manufacturers portfolio in the world. They are used in all vehicle types: from premium vehicles to even the smallest sedan platforms and from vans up to heavy trucks and travel busses. With the introduction of the collision prevention assist®, radar sensors have become even standard equipment in passenger cars [1]. The major reason for the success story of automotive radar is its physical principle that offers unique performance features at reasonable costs. Among others, they are independent from environ mental factors like illumination and weather conditions, are capable of

T. Bertram (Hrsg.): *Fahrerassistenzsysteme 2018,* Proceedings, S. 1–17, 2019.
https://doi.org/10.1007/978-3-658-23751-6_1

directly measuring positions, material parameters and Doppler velocities, have multiple field of views and are easily compatible to vehicle designs and paintings. The packaging benefit allows a 360° coverage from the vicinity to extreme far ranges. Radar performs under conditions where other sensor types fail or suffer in performance and it can virtually look through vehicles by exploiting reflections between the road surface and the vehicle's underbody and hence makes the invisible visible. Very often, the radar signal is the only remaining signal out of the total perception system the controller sees in ad verse environment conditions or in the dark.

Over decades, the performance requirements increased steadily from simple detection and ranging tasks in blind spot monitoring or cruise control systems up to smart environment perception tasks for present day semi-autonomous evasion and braking functions [2]. However, the utmost push in performance requirement is initiated with the trend towards highly automated driving and fully autonomous driving [3–6]. In 2013, the first stride ahead towards higher automation has been made with the fully autonomous Bertha drive of a Mercedes-Benz research sedan [3–6]. The major lesson learnt was that future automotive radar systems have to provide imaging like capabilities, i.e. very high resolution in space and velocity, and have to interact in radar networks composed of four or more short-, mid-, and far range radars, which allow for 360° highly comprehensive perception tasks. This insight is backed by the published sensor set-ups of other groups developing driverless vehicles [7–9]: a from year to year steadily increased number of radar sensors in their setups can be noticed.

1.2 Paradigm-Shift in Development Guidelines: Deep Learning and Model-Based Approaches Together on Imaging Like Radar Data

In general, sensors and perception algorithms for driver assistance and active safety systems are designed to observe the direct neighbours of the ego vehicle and to identify the relevant objects. They are designed for specific tasks, where human drivers often struggle: these are either situations where constant awareness and quick reactions are necessary or where the driver has to estimate distances and velocities accurately. Example applications are lane keeping, emergency braking and parking assistance systems. However, human drivers excel any of today's assistance systems in tasks, where contextual understanding of the surroundings is needed, e.g. the anticipation of other road user's driving behaviour, reaction to lost cargo or path planning in construction sites. Driverless cars have to cover all activity areas and ideally under all conditions. Thus, sensors and perception algorithms for driverless driving vehicles face a much harder task. They have to describe the complete static and dynamic neighbourhood of the ego-vehicle and have to understand the interplay of both. Considering the natural benefits of radar lined out above, it is logical that on a global scale radar becomes the back-bone of driverless vehicles. Therefore, radar sensors and their perception algorithms have to be further developed from a pure detection and ranging system, to a complete semantic environment understanding system [10–12]. One important consequence is that radar signal processing has to be complemented with machine learning and scene understanding concepts to keep radar in the leading edge of remote sensing. A pre-requisite for applying new algorithm concepts is that radar data becomes imaging like, in order to provide many valuable detections per object. Hence, future radars have to provide high resolution and accuracy in space and velocity [10, 12, 13]. Radar systems developed for level 1 to 3 driving tasks will fail in providing the required information [14].

1.3 Is Deep Learning Really the Answer to Everything?

The intensive and wide spread use of deep learning methods for classification and other tasks in the automotive area has been initiated by their successes in image recognition tasks. It was pushed by various companies like Nvidia, who provide new automotive chips and powerful GPUs, Google, Microsoft and Apple, who created deep learning software packages and sophisticated algorithms. Until now, radar took advantage by exploiting its unique characteristics, gaining the required information by model based approaches using well-known physical relationships. Although a lot has been achieved using those techniques, the success that has been reported using deep learning requires radar developers to adapt to new methods in order to push their achievements even further.

End to end approaches exist which ignore all signal processing steps, but they require huge amounts of labelled training data and computing power. The main disadvantage of deep learning algorithms is that it is very hard to determine their behaviour in unknown situations, which are not present in the training or test dataset. Therefore, if a model-based solution to a problem is well known, this model is for us almost always preferable to a deep learning method. For example, it seems unlikely that well established signal processing methods like Fourier transforms or peak detection algorithms will be replaced by Deep Learning, because a clear task and a working method are already at hand.

Then again, on some real world problems it is virtually impossible to determine a good model. For example, humans can easily classify objects as a car, but it is difficult to describe the classification task in a model. Variety in colour, the number of visible wheels, occlusion or even a cover with a tarpaulin exemplify the difficulty of the task.

For these tasks, deep learning can help to achieve the required performance, because the generated model is so complex that it cannot be designed by a human. It seems therefore logical that traditional and deep learning techniques should cooperate, to enhance their specific strengths.

Unfortunately, an easy adoption of all available techniques and tooling ecosystem fails due to the totally different nature of radar data compared to vision systems. Hence, research activities concentrate among other things on getting the right data representation to benefit from the available ecosystem and to apply radar data to deep learning techniques.

The first group that applied deep learning to radar to solve the problems of semantic classification of static and dynamic objects as well as the estimation of the course of the road using radar sensors, was Daimler research [15–20].

The class information of dynamic objects can be used to choose an object-specific tracking method. This is necessary since the motion pattern of e.g. pedestrians, cyclists, cars or trucks differs significantly and therefore custom motion models are essential. If mirror effects, ghost targets and other sensor artefacts are annotated correctly as such and algorithms further down the processing chain can rely on semantic class labels for each measured reflection, all these methods become much easier and for example tracking algorithms can perform better and faster.

The fusion of static and dynamic object classification generates a thorough understanding of the traffic scenario and therefore critical situations can be anticipated at an early stage and vulnerable road users can be considered with particular caution. In addition to classification, deep learning makes also road course detection possible, so that for example even on snowy roads path planning with radar data is possible.

Classification of static objects improves landmark detection and validation which in turn optimizes the self-localization and allows driverless parking even in parking garages [21, 22]. For example, movable objects like parked cars or trash cans are rejected as possible landmarks so that only truly static objects are used for localization. Semantic information of static objects can also be utilized for associating the radar landmarks with landmarks from third party maps and thereby localizing the ego vehicle on nonradar specific maps.

The subsequent three sections will provide examples which underline the potential of artificial intelligence and deep learning in model based concepts for radar perception. The Sect. 2 focuses on radar applications for dynamic objects and give detailed information on how moving objects can be semantically classified and how this information can enhance tracking algorithms. In the Sect. 3, we turn to the static world: here we introduce semantic radar grids and their application in classification and road course estimation. Finally, we give an outlook to high resolution algorithms which motivate the development of high resolution radar to enhance the aforementioned applications.

2 Radar Application for Dynamic Objects

2.1 Classification

The eponymous applications for radar – detection and ranging – allows to make basic statements like: "In distance x is an object". If additionally Doppler velocity, azimuth angle and reflected energy can be obtained from the sensor, the claim can be formulated: "In distance x and direction ϕ is an object with radial velocity v and that objected reflected the energy amount E". However, the nature of the detected object, that is the *semantic class* the object belongs to, remains unknown. The aim of classification algorithms is to specify the "what" of the detection: Does a measured reflection belong to a moving pedestrian, a bicyclist, a truck or a car?

Semantic knowledge about the detection can be utilized in many scenarios: class specific tracking algorithms can easily outperform general tracking approaches, path planning for autonomous vehicles has to take information about the surroundings into account, and critical situations may be reduced or even avoided if other road users are identified early on.

Only recently, this new application of radar sensors became more popular and gained attention especially in the automotive field [18, 23]. This timing is not a coincidence but tightly coupled to the rapid improvements in the fields of machine learning and image understanding. The underlying methods and algorithms – which were originally designed for image-like data – can also be translated to the radar world. This translation step, i.e., the modification of the algorithms which work seamlessly on camera images to radar data, is by no means trivial or straight-forward and is still ongoing research.

The difficulties already start at the labelling stage: Almost all machine learning approaches rely on manually annotated data, so called "labelled data". During a training stage, neural networks or other machine learning algorithms learn from this annotated data and are then able to apply this learned knowledge to new unseen data. For images, large publicly available databases with millions of annotated samples exist and it is therefore easy to start with the optimization or development of new algorithms. For radar data, no such free dataset exists. The reasons for the lack of publicly available annotated radar data are manifold. Firstly, radar data with ego-motion compensated Doppler velocities is harder to obtain than camera images, which everyone can capture with their smartphones. Secondly, one has to decide on which data level the labelling should be performed: after receiving the reflected wave, various pre-processing steps like fast Fourier transforms and peak detection algorithms are applied and for each intermediate representation of the data, arguments can be found for and against labelling at this stage. Thirdly, the way radar sensors perceive the world is very different from the way humans do. Therefore, human annotators have to be trained to work on radar data so that they understand some of the sensor properties. Experience showed that humans can easily identify traffic signs in a camera image but struggle to identify radar reflections at given distances and azimuth angles that belong to the same object. In addition to these challenges, misalignments of the sensors, time synchronization problems with the ego-motion, multipath reflections, mirror effects and ghost measurements increase the difficulty of the labelling task.

To achieve our aim to classify measured radar reflections, we developed our own labelling software, collected data, defined a class hierarchy and labelling instructions, and trained labelling experts who annotate the data. Our current dataset for classification of dynamic objects contains measurements from over 4.5 h of driving, that is, over 100 million radar reflections were collected of which 3 million belong to 6200 different moving objects. For each of the 100 million reflections an annotator had to decide if said reflection belongs to a dynamic object and if so, to which class this object belongs to.

To use the labelled data for training of our deep learning algorithms, some pre-processing is necessary. In a first step, the clustering, all radar reflections that belong to the same object are grouped together. The formed clusters are then the basis for the feature extraction. In this second step, statistic properties and other descriptive quantities are calculated over all reflections in a cluster. The selection of meaningful features is a non-trivial task since it can be easily proven that the approach "the more features the better" is actually harmful for the classification scores. At the same time, one has to keep in mind that manually created features have only a certain ability to generalize to unseen data.

Our current approach is depicted in Fig. 1. We use a special kind of neural network, a so called long short-term memory network (LSTM), which is able to learn from a sequence of input data and can hence take the temporal evolution of an object into account. These kind of networks are also used by popular devices such as Amazon's Alexa.

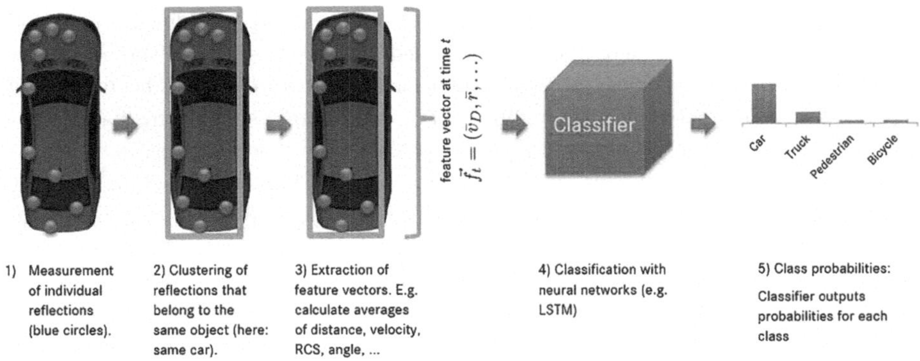

Fig. 1. Classification pipeline. From left to right: measurement of targets, clustering, feature extraction and application of classification algorithms.

One may wonder why machine learning methods are necessary at all for the classification tasks and whether "classic" algorithms which do not rely on millions of labelled reflections would not work just as good. In Fig. 2, different examples of reflections that belong to individual cars are displayed. One can easily notice that shape, extension and distribution of the reflections differ greatly among the examples. Additionally, one can find clusters of pedestrian groups whose distribution of reflections looks similar to those of a single car. Simple algorithms that only target the shape of a cluster and the measured Doppler velocities cannot generalize well enough to deliver reasonable classification results without creating many false positives. The complicated decision rules that are generated by a deep neural network cannot be easily (re-)produced by "classic" algorithms and history has shown that deep learning excelled all other approaches in this kind of tasks.

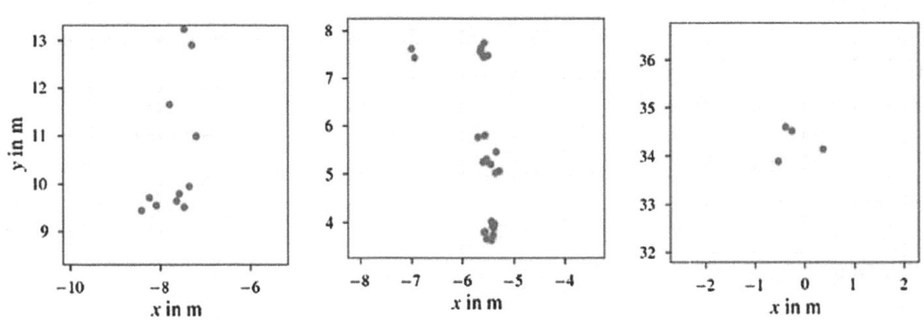

Fig. 2. Three real world examples of radar reflections from three different cars. Data from one sensor, accumulated over two cycles, is shown. With greater radial distance, the number of measured reflections decreases and hence the classification accuracy also declines.

The last ten years in the field of scene understanding with camera images displayed impressively that a classifier that generates its own features outperforms any algorithm that can only work on a pre-defined feature set. It seems therefore desirable to build classifiers that take the raw radar data as input and create their own descriptive features with which the classification can be done. This step is still ongoing research but first experiments at Daimler research showed promising results so that semantic understanding of moving objects using only radar is in reachable distance.

2.2 Extended Target Tracking

Target tracking algorithms play a vital role in the fusion and behaviour planning activities in the autonomous driving functions. Earlier advanced driver assistance systems and some limited autonomous functions [5] have managed to carry out their safety critical functions by performing point based target tracking. The target extensions such as length and width are approximated heuristically by the algorithm since automotive sensors of those days did not had sufficient resolution. As the resolution of the sensors increased in parallel to the affordability in terms of purchase cost, this factor enables us to improve the target tracking applications w.r.t estimation of target size and shape, yaw-rate and orientation. This allows us to program and model higher autonomy driving functions. Implementing a real-time automotive extended target tracking algorithm [24] is very closely tied with the availability of computing resources, communication band-width and allowed operational latency in the processing chain.

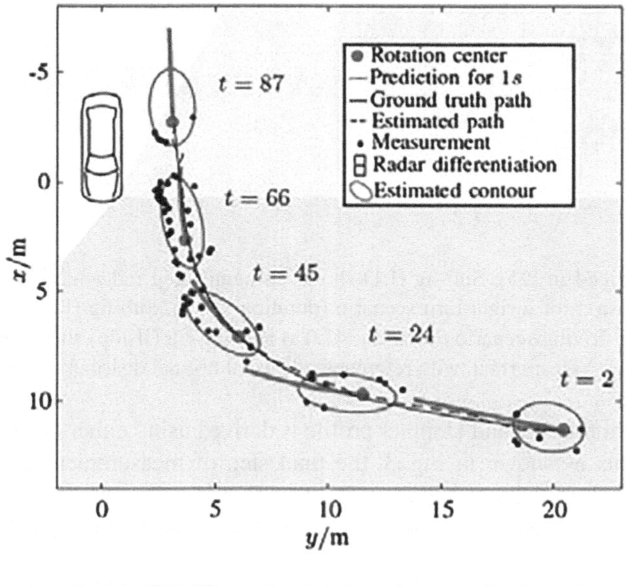

Fig. 3. As described in [25]. The estimated shape is approximated using an Ellipse.

It is vital to understand the important use-cases that necessitate a robust and stable sensor-set specific extended target tracking algorithm: all overtaking manoeuvres, collision avoidance and tight corner manoeuvres, tracking of group of pedestrians, bicyclists, targets of unidentified type and form.

Before we go ahead with discussion of estimating the size of the target under observation, it is noteworthy to consider parameters such as the velocities in x and y direction, as well as the yaw rate: (x·, y·, ‘-·). These parameters are estimated [25] by building knowledge about the Doppler velocity distribution of the different observation across the target vehicle. The higher the radar resolution is in space and velocity, the better and more robust the Doppler profile can be deduced and the yaw rate calculated. Prior shape models play an important role in robust estimation of shape of the target. As shown in Fig. 4 and as derived in [26], a prior spatial model along with Doppler profile called Volcanormal model was proposed. This model uses a symmetric positive definite 2-d Ellipse to describe the size of the target under observation. The estimated Doppler profile is used to determine the yaw rate ‘-· which enables us to predict the orientation of the target under observation.

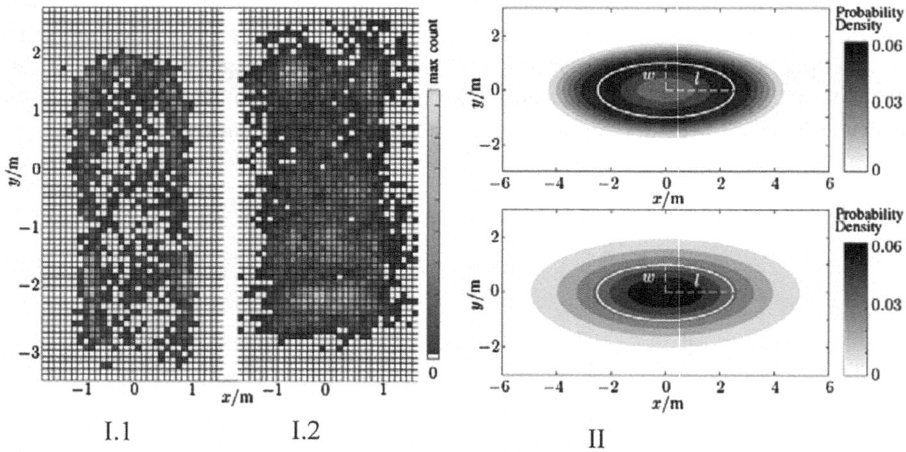

Fig. 4. As described in [25]. Sub-fig (I.1) shows an aggregated radar measurement grid with 0.1 m cell resolution for a right-turn scenario (duration: 7.6 s). Sub-fig (I.2) displays a grid of a different urban driving scenario (duration: 47.0 s) and sub-fig (II-top) shows our radar spatial distribution, called Volcanormal, with reference to a usual normal distribution (II-bottom).

Once the prior model and Doppler profile is derived using either one or more frames of measurements as shown in Fig. 3, the final step of measurement differentiation is to calculate the measurement delta's, a Doppler-fit and the shape or the contour of the target under observation is estimated based on the measurement spread and ‘-·. The parameters length and width can be extracted using the Ellipsoid contour information of the observed target. The effectiveness of this approach is displayed in more detail in [26].

2.3 Classification Assisted Tracking

Classification assisted tracking (CAT) uses the information obtained from the classification algorithms as an important source in the information gain likelihood calculation in addition to the kinematic information of the measurement and state space. This enables us to perform robust tracking in a dense target and clutter zone. Target separation ability of the tracking algorithm is enhanced to a very high success rate. This subject is discussed in detail in [27] and [28]. CAT uses traditional Bayesian probabilistic framework with different hypothesis and combinatorics nodes to derive its different likelihoods. However, a numerical approximation suitable for a real-time implementation is highly possible.

3 Scene Understanding of Non-moving Targets

Only until recently, the static world, i.e. reflections from non-moving targets, were solely considered as obstacles – if taken into account at all

With the demand for more advanced driver assistance systems, a better perception of the world is needed and hence also semantic information about non-moving objects is necessary.

Although radar data is sparse in comparison to LIDAR or camera, a lot of information can be inferred if it is accumulated over multiple times steps and radar physics is exploited. To this end, occupancy grid maps specifically designed for radar data can be used.

In Fig. 5 an occupancy grid can be seen, which depicts the static objects around the ego-vehicle. Probability of occupancy of individual cells is colour coded with yellow depicting a high probability and blue the state of free space.

Fig. 5. Occupancy grid map centered around the ego-vehicle.

3.1 Semantic Radar Grids

Semantic knowledge enables us humans to act appropriately. We can find a proper parking spot, even if no road markings are present. We know that we have to drive carefully in areas where parked cars block the view to the side walk where children are playing. We identify our position on proper landmarks, like buildings, trees and the road course and do not take parked cars and other movable objects into account.

For dynamic objects, the distribution of the Doppler velocity is a characteristic radar feature, to distinguish between different object classes. This cannot be used for the static world. Therefore, the distribution of the positions of measurements and the intensity of the backscattered signal are key feature elements, which are accumulated in the radar grids [15–17, 20].

Deep learning methods from image processing, as used for example in the "Large Scale Visual Recognition Challenge", can be adopted to extract semantic information in radar grids. Deep convolutional neural networks (CNNs) have been proven capable to learn models for a variety of different data from many fields. Hence, using the right preprocessing steps, this technique can also be used for radar. One appealing property of this approach is that not only information of one object is taken into account but also the object's surroundings are considered.

Even though objects have a very different appearance in radar than in images and features we are familiar with are missing, e.g. color, a lot of advantageous properties can be used to get the right classification result. Because distances are measured directly, the radar sensor does not suffer from perspective distortion, so that the real size of an object can be determined. Additional material properties can be detected by measuring the radar cross-section. This very different perception of the world is unique to radar and hence situations can be resolved in which cameras are fooled by mirrors, monitors or printed images.

For the training of deep learning algorithms, an annotated dataset is needed. Just as in the case of radar data for dynamic objects, no public dataset is available and therefore we created our own dataset.

To build up the dataset efficiently, the following labeling process is pursued: data is collected in a grid whose cell positions are given in a world coordinate system and hence reflections from static objects, measured at different times, fall into the same area in the grid, irrespective of the motion of the ego-vehicle. It is then possible to assign class labels to regions in the grid and thereby labeling all reflections of the same object at once. With this approach, over 50.000 real world objects were labeled, yielding several millions of training examples from different measurement times.

We proved that classification of different object classes like cars curbstones, trees, poles and plants is possible. Because grids – other than most images – have no natural orientation, all rotations of a grid have to be learned [15–17, 20]. An example of a semantic grid map in which each pixel is assigned to a class is shown in Fig. 6.

Fig. 6. Semantic radar grid map. Same scene as in previous figure. Colors indicate class affiliation (Color figure online)

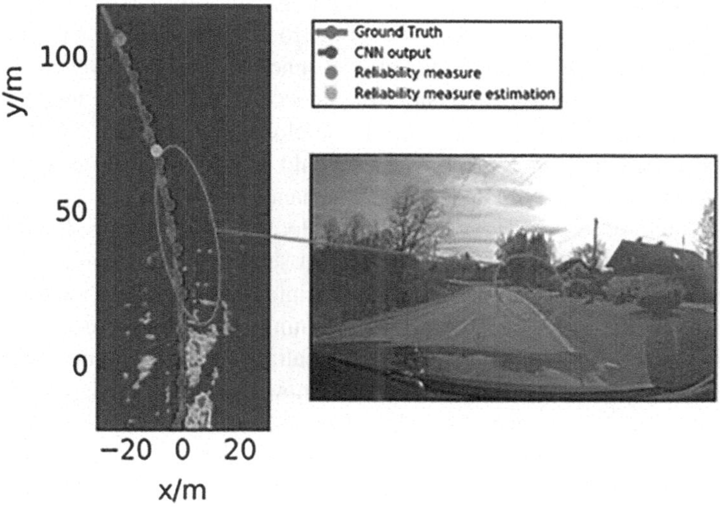

Fig. 7. Example of road course estimation. Front facing documentation camera on the right, occupancy grid with label, road course estimation, reliability measure and its estimation on the left. (Color figure online)

3.2 Road Course Estimation

One of the essential tasks for an autonomous vehicle is to gather knowledge about the road path ahead. With this information, it is possible to plan the path of the future trajectory and to verify a digital map. Radar sensors are especially useful for this purpose because of their ability to detect targets even in high distances which is needed to capture the geometry of the road.

Radar sensors mounted on a vehicle are unable to receive reflections from the road or lane markings directly, and hence the course of the road has to be found indirectly through the surrounding areas. Objects like guard rails, reflector posts or even the transition from tarmac to the road's surroundings (e.g. grass on the side of country roads) can be used to draw conclusions on the most likely road course. Since the reflections from these objects are relatively sparse, especially from the edge of the road, it is beneficial for further processing to transform the received radar data into occupancy grids to accumulate the radar targets and filter them in the time domain.

In the literature, the road course is most commonly estimated by applying rule based algorithms on occupancy grids [29]. These algorithms are built on a few assumptions and mostly on model based road courses. The main assumption is that the road border objects are parallel to the road course while the most commonly used road model is that of a clothoid or an approximation of a clothoid through a cubic polynomial in lateral direction. Yet these assumptions don't always hold true. The road border is not always parallel to the course of the road or the lane the car wants to stay in. This assumption is for example violated at junctions and highway exits. Also the assumed model of the road course isn't always accurate. At the beginning of a curve on a rural road, there is a transition from one clothoid to another which cannot be approximated accurately by a single one. It would be necessary to detect these cases where the assumptions are violated and then be handled separately.

Since this approach is not very practical, we looked for a way to solve the problem with a single procedure. As described in the last section, radar occupancy grid maps allow us to transform the measured reflections into a shape which modern convolutional networks can take as an input [20]. Training data for road course estimation can be easily generated with a high precision localization system. With such a system, the position of the lane the car is driving in is known in the occupancy grid and thus, labelled data is created automatically while driving.

For the use in an automated application, not only the road course has to be estimated, but it also has to be known, how reliable this estimate is. The use of CNNs shows another advantage in this regard, since such a reliability measure can be trained simultaneously with the regression. To train the neural network, we recorded over 700 km of rural roads and highways in southern Germany. Figure 7 shows an example of the test set of a rural road with a junction. Classical algorithms will fail in such a case because of the large scattering area on the right hand side (see red markings in the figure) of the road which leads to a not recognizable road border. Neural networks are able to detect the specific conditions in a scene and react accordingly.

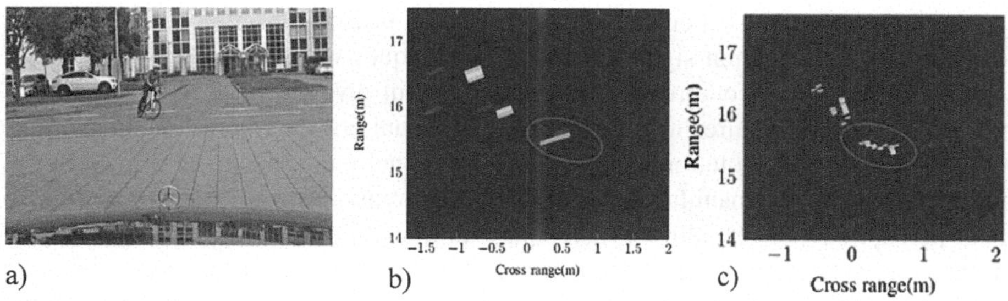

Fig. 8. (a) Camera view of traffic scene. (b) Measured reflections without high resolution. Only few reflections of the bicycle are visible. (c) Highly resolved radar image of the cyclist. Notice that the tilted front wheel of the bicycle is also clearly visible.

4 Future Trend: High Resolution Radar Combined with Artificial Intelligence

The resolution of a radar in a given dimension defines its ability to distinguish between two targets separated by a minimum value in the same dimension. Automotive radars typically measure three different parameters for every target where these parameters are often referred to as dimensions: range, Doppler velocity and angle. For instance, a resolution of 1 m in the range dimension would imply that the radar can distinguish between two targets that are separated by at least a distance of 1 m. If this distance is reduced to any value below the resolution limit, the radar will simply see only a single target. The resolution limit of a radar is determined by its hardware specifications. For example, the amount of time that a radar illuminates a target, determines its Doppler resolution. This means, the longer a moving target is illuminated by the radar signal, the more exact its Doppler velocity components can be estimated.

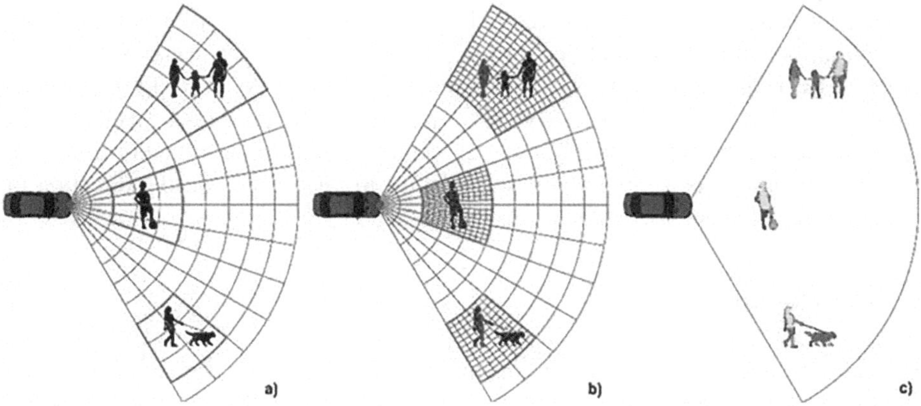

Fig. 9. Two-stage classification algorithm: (a) Stage one identifies regions of interest in regular resolution FoV. (b) The second stage uses high resolution processing on the RoIs for a final object classification as depicted in (c).

A viable alternative to enhancing the hardware in order to improve resolution is to employ high resolution signal processing techniques without modifying the hardware. Like any other compromise, there is a certain give and take. High resolution processing usually requires much more computational power, whereas modifying the hardware for better resolution can keep computational requirements at a lower level, however it increases manufacturing costs. For a transition from radar detection to radar perception, high resolution radars play an integral role. With the availability of highly resolved targets in day-to-day traffic scenarios the task of target classification and identification can be performed with a much higher level of certainty, making the safety of all road users feasible.

As previously mentioned, high resolution algorithms require an increased amount of computational power. A possible approach for solving this problem is to apply these algorithms only within certain regions of the entire traffic scene. This method is often referred to as region of interest (RoI) or frequency selective processing. Figure 8 shows the radar images obtained from an exemplary traffic scenario. The plot in the middle shows data that have been processed using conventional low complexity processing as compared to the plot on the right which has been obtained with the use of high resolution techniques.

To explain this concept, imagine a traffic scenario where pedestrians are walking in a crowded scenario on their own, in groups, or with pets. A typical mid-range automotive radar will be able to see up to 70 m in range and has field of view (FoV) of over 100°. In such a scenario where detection and classification of pedestrians is given top priority, it suffices to apply high resolution for limited ranges (e.g. 10–15 m) and a smaller FoV. The remaining sections of the traffic scenario can be resolved by means of conventional processing with a lower resolution. The incorporation of RoI processing in the classification process is done in a two-step algorithm that uses regular resolution for a classification of RoIs as a first step. In a second step, the highly resolved RoIs are assigned an object class (cf. Fig. 9). The higher resolution enables the classifier to achieve very accurate classifications results; it is even possible to identify different body parts of a pedestrian. This is e.g. crucial for the prediction of gait starting or stopping.

The availability of highly resolved radar images opens the door on further classification and prediction possibilities. In Fig. 8c, the bicycle in the scene is processed using a HR technique [30]. If only low resolution data is available, very few reflections of the bicycle can be measured. However, in the HR processed image, the tilt of the front wheel of the bicycle is clearly exhibited and this information can used as input for image processing algorithms that try to predict the turning direction of cyclist based on wheel orientation.

5 Conclusion

The paper outlined the challenges for automotive radar to provide the required perception information for autonomous vehicles. It has been outlined that Radar data processing toolchains have to be revolutionized by applying artificial intelligence and advanced model based signal processing in a synergetic manner. From the analysis it became obvious that future automotive radars have to provide extreme resolution in space and velocity.

References

1. Daimler AG: Collision prevention assist. https://www.mercedes-benz.com/de/mercedes-benz/innovation/mit-abstand-am-besten-assistenzsysteme-von-mercedes-benz/. Zugegriffen: 26 Februar 2018
2. Daimler AG: Mercedes-Benz-Intelligent-Drive. https://www.mercedes-benz.com/en/mercedes-benz/innovation/mercedes-benz-intelligent-drive/. Zugegriffen: 26 Februar 2018
3. Daimler AG: Bertha-Benz Drive. http://media.daimler.com/marsMediaSite/de/instance/ko/Pionierleistung-Autonome-Langstreckenfahrt-im-Ueberland–und-Stadtverkehr-Mercedes-Benz-S-Klasse-INTELLIGENT-DRIVE-faehrt-autonom-auf-den-Spuren-von-Bertha-Benz.xhtml?oid=9904223. Zugegriffen: 26 Februar 2018
4. Dickmann, J., Appenrodt, N., Brenk, C.: Bertha fährt autonom. In: Automobil Elektronik, pp. 44–47, March 2014
5. Dickmann, J., Appenrodt, N., Klappstein, J., Bloecher, H.L., Muntzinger, M., Sailer, A., Hahn, M., Brenk, C.: Making Bertha see even ore: radar contribution. IEEE Access **3**, 1233–1247 (2015)
6. Dickmann, J., Appenrodt, N., Brenk, C.: Making Bertha See. In: IEEE Spectrum, pp. 40–46, August 2014
7. Waymo: Waymo sensor suite. https://cdn-images1.medium.com/max/1600/1*9RQQY5ehT-W5dYAhwdU4Pdg.jpeg. Zugegriffen: 26 Februar 2018
8. Uber: Self-Driving-Uber-Vehicles-San-Francisco. http://www.rideshareconnection.com/self-driving-uber-vehicles-san-francisco/. Zugegriffen: 26 Februar 2018
9. General Motors: General motors sensor suite. http://www.repairerdrivennews.com/wp-content/uploads/2017/12/general-motors-av-autonomous-sensors57-2-1024x576.jpg. Zugegriffen: 26 Februar 2018
10. Dickmann, J.: Automotive radar and radar based perception for driver-less cars. In: Ben Gurion University Radar Symposium. Ben-Gurion University of the Negev, Israel (2017)
11. Dickmann, J.: When will we give up driving our cars? A radar developers point of view. In: ISSCC, San Francisco (2017)
12. Dickmann, J.: Automotive radar systems at daimler: past-present-future. In: IEEE Radar Conference, Enabling Technologies for Advances in Radar, Philadelphia (2016)
13. McCloskey, D.: Radars as a complementary sensor for autonomous driving at Waymo. In: 14th European Radar Conference on EURAD-Workshop: Future Automotive Radar Systems, Nurenberg (2017)
14. Wyholt, A.: Scalability of sensor systems for active safety and autonomous functions. In: International VDI Conference – Automotive Sensor Systems, München (2018)
15. Lombacher, J., Hahn, M., Dickmann, J., Wöhler, C.: Detection of arbitrarily rotated parked cars based on radar sensors. In: 16th International Radar Symposium (IRS) (2015)
16. Lombacher, J., Hahn, M., Dickmann, J., Wöhler, C.: Object classification in radar using ensemble methods. In: IEEE MTT-S International Conference on Microwaves for Intelligent Mobility (ICMIM) (2017)
17. Lombacher, J., Hahn, M., Dickmann, J., Wöhler, C.: Semantic radar grids. In: IEEE Intelligent Vehicles Symposium (2017)
18. Schumann, O., Hahn, M., Dickmann, J., Wöhler, C.: Comparison of random forest and long short-term memory network performances in classification tasks using radar. In: Sensor Data Fusion: Trends, Solutions, Applications (SDF), Bonn (2017)
19. Giese, T., Klappstein, J., Dickmann, J., Wöhler, C.: Road course estimation using deep learning on radar data. In: International Radar Symposium (IRS 2017), Prag (2017)

20. Lombacher, J., Hahn, M., Dickmann, J., Wöhler, C.: Potential of radar for static object classification using deep learning methods. In: IEEE MTT-S International Conference on Microwaves for Intelligent Mobility (ICMIM) (2016)
21. Werber, K., Klappstein, J., Dickmann, J., Waldschmidt, C.: Interesting areas in radar gridmaps for vehicle self-localization. In: International Conference on Microwaves for Intelligent Mobility (ICMIM 2016), San Diego (2016)
22. AdaptIVe Project: AdaptIVe – final project results (2017)
23. Schumann, O., Hahn, M., Dickmann, J., Wöhler, C.: Supervised clustering for radar applications: on the way to radar instance segmentation. In: ICMIM – (2018)
24. Duraisamy, B., et al.: Track level fusion of extended objects from heterogeneous sensors. In: 19th International Conference on Information Fusion (FUSION) (2016)
25. Kellner, D., et al.: Tracking of extended objects with high-resolution doppler radar. IEEE Trans. Intell. Transp. Syst. **17**(5)1341–1353 (2016)
26. Brosseit, P., et al.: The volcanormal density for radar-based extended target tracking. In: IEEE 20th International Conference on Intelligent Transportation Systems (ITSC) (2017)
27. Duraisamy, B., et al.: Combi-Tor: track-to-track association framework for automotive sensor fusion. In: IEEE 18th International Conference on Intelligent Transportation Systems, Las Palmas (2015)
28. Duraisamy, B., et al.: Track level fusion algorithms for automotive safety applications. In: International Conference on Signal Processing, Image Processing & Pattern Recognition (2013)
29. Sarholz, F., et al.: Evaluation of different quality functions for road course estimation using imaging radar. In: Intelligent Vehicles Symposium (IV) (2011)
30. Dehkordi, S., Appenrodt, N., Dickmann, J., Waldschmidt, C.: Region of interest based adaptive high resolution parameter estimation with applications in automotive radar. In: IRS, Bonn (2018)
31. Duraisamy, B., et al.: Influence of the sensor local track covariance on the track-to-track sensor fusion. In: IEEE 18th International Conference on Intelligent Transportation Systems (2015)
32. Duraisamy, B., et al.: Object management strategy for an unified high level automotive sensor fusion framework. In: IEEE International Conference on Multisensor Fusion and Integration for Intelligent Systems (MFI), Baden-Baden (2016)
33. Mayer, W.: Abbildender Radarsensor mit sendeseitig geschalteter Gruppenantenne, Dissertation. Institut fuer Mikrowellentechnik, University of Ulm, Ulm (2008)
34. Daimler AG: Remote parking pilot. http://media.daimler.com/marsMediaSite/de/instance/ko/Remote-Park-Pilot-Ferngesteuertes-Parken-per-Smartphone-App.xhtml?oid=9361355. Zugegriffen: 26 Februar 2018
35. European Project MOSARIM: MOre Safety for All by radar Interference Mitigation, Proj. Ref. No: 248231, FP7-ICT (2014)
36. Muntzinger, M., Aeberhard, M., Zuther, S., Schmid, M., Dickmann, J., Dietmayer, K.: Reliable automotive pre-crash system with out-of-sequence measurement processing. In: IEEE Intelligent Vehicles Symposium, pp. 1022–1027 (2010)
37. Andres, M., Feil, P., Menzel, W.: 3D-scattering center detection of automotive targets using 77 GHz UWB radar sensors. In: EuCAP, Prag (2012)
38. Hosseini, A., Diewald, F., Klappstein, J., Dickmann, J., Neumann, H.: Modification of the landweber method based on the conjugate gradient method to restore automotive radar images. In: International Conference on Systems, Signals and Image Processing (IWSSIP), Wien (2012)
39. Kellner, D., Barjenbruch, M., Klappstein, J., Dickmann, J., Dietmayer, K.: Instantaneous full-motion estimation of arbitrary objects using dual doppler radar. In: Intelligent Vehicle Symposium (IV 2014), Dearborn (2014)

40. Sarholz, F., Mehnert, J., Klappstein, J., Dickmann, J., Radig, B.: Evaluation of different approaches for road course estimation using imaging radar. In: Intelligent Robots and Systems (2011)
41. Dickmann, J., Klappstein, J., Hahn, M., Muntzinger, M., Appenrodt, N., Brenk, C., Sailer, A.: Present research activities and future requirements on automotive radar from a car manufacturer's point of view. In: IEEE MTT-S International Conference on Microwaves for Intelligent Mobility (ICMIM), Heidelberg (2015)
42. Dickmann, J., Meinel, H.: Automotive radar: from its origins to future directions. MWJournal **56**(9), 24–40 (2013)

Improving the Environment Model for Highly Automated Driving by Extending the Sensor Range

Nicole Beringer(✉)

Elektrobit Automotive GmbH, Am Wolfsmantel 46, 91058 Erlangen, Germany
Nicole.Beringer@elektrobit.com

Abstract. This paper describes a novel approach to cope with driving scenarios in highly automated driving which are currently solved only by the driver's control. The approach presented in this paper is currently being implemented as a prototype to be used in our test fleet. It combines techniques well established in robotics like Simultaneous Localization And Mapping (SLAM) as well as end-to-end protection and image compression algorithms with big data technology used in a connected car context. This allows enhancing the positioning of individual vehicles in their Local Environment Model (LEM). This is the next step to overcome current dependencies to in-vehicle sensors by using additional cloud-based sensor processing to gain information.

1 Introduction

In highly automated driving (HAD) there are still a lot of driving scenarios where the driver needs to take over control. The reasons for taking over vary from limitations of the range of ego sensors or recognition algorithms to required information, e.g. infrastructure information like traffic lights, which cannot be derived from in-vehicle sensor observations. What all have in common is that any reaction, from the driver as well as from a driver assistance feature, needs to be in time. This becomes clear when looking at the range of ego sensors (e.g. LiDAR sensors about 40 m ahead). The driver may want to have the speed reduced in advance before a speed sign is reached or be warned in time to take over control if the HAD road ends. It is quite clear that it needs more than just a high quality in-vehicle sensor processing in order to obtain a wide range of HD information needed for HAD.

Imagine the following scenario: your car is equipped with all necessary sensors to allow HAD. Your car drives on a motorway which is suitable for HAD while you are asleep. During an overtaking manoeuver, the sensors which are needed for HAD stop working due to some technical defect and you cannot take over. Fortunately, the car is able to bring itself into a safe state. But even if the car manages to navigate to the

N. Beringer—Program Manager Highly Automated Driving.

T. Bertram (Hrsg.): *Fahrerassistenzsysteme 2018*, Proceedings, S. 18–29, 2019.
https://doi.org/10.1007/978-3-658-23751-6_2

emergency lane for the safe state it still leaves the passengers in danger. That is where our approach enters the game: By fusing the environment model of our endangered vehicle in the described scenario with other environment information, it can reach a safe state which is also not dangerous for the passengers: For instance, by driving to the next parking lot, even if it is several kilometers away from the position where the sensors of the car initially stopped working.

The next section gives an overview of current approaches. In Sects. 3 and 4 the underlying algorithms are described. Section 5 presents our current results. The last section gives an outlook for future work.

2 Current Approaches

Within recent years, the trend shifts towards obtaining a wide range of information for HAD from a cloud-based solution instead of only relying on a single-source evaluation of ego sensors, even if the in-vehicle sensor processing is of high quality. In the literature you can find several publications, e.g. [1–8] as well as webinars (e.g. [9–11]) dealing with possible solutions.

One major idea is to enrich existing navigation maps with HD information as described in [12] and [13]. This is quite an efficient method to provide highly accurate, up-to-date maps for the development of self-driving vehicles based on map data information in the electronic horizon. For further details on the electronic horizon, please take a look at one of the following references [14–16].

The described approaches need to meet the quality criteria as follows:

- Up-to-date-ness: Data is continuously transferred to the cloud. Any changes to the road for instance are detected as soon as the vehicle passes by. The sensor quality may need to be confirmed by other vehicles.
- Reliability: Deviations of map contents used in the vehicle are detected by the vehicle sensors and are directly submitted to the cloud. The data is always up-to-date and with this, the probability of unconscious changes is drastically reduced as other vehicles always crosscheck the map content as well.
- Precision: The map becomes more precise with more collected data.

Limitations of ego-sensor ranges and recognition algorithms can be overcome well. Using classic navigation maps as base layer, HD features can be layered on top.

Grid-based approaches as in [17–20] need to have a general idea about the structure of an HAD system with its components that are required for a dedicated functionality. The current approaches follow the sense-plan-act-principle [21]. To reduce complexity, sensors, components with special tasks (e.g. control), and actuators need to appropriately interact within an HAD system.

Neumann et al. [22] presented an industrialized version for free grids based on stereo camera input.

3 Simultaneous Localization And Mapping (SLAM)

Our approach for determining the free and occupied areas (occupancy grids [23, 24]) in the environment of a vehicle – the Local Environment Model (LEM) – benefits from a very popular algorithm in robotics, the Simultaneous Localization And Mapping (SLAM) algorithm, also known as Concurrent Mapping and Localization (CML).

In the literature several approaches are available to overcome errors:

The authors of [25] present a local map-joining method for indoor applications based on building independent local maps. With their approach, they efficiently show that they can significantly reduce linearization errors which usually lead to inconsistent estimates well before the computational problems arise.

In [26], the focus is on reducing computational complexity in large environments for outdoor applications using a Hybrid Metric Map (HYMM). The work covers the following most important challenges: map representation, consistency, and data association.

For them, successful data association involves association of correct measurement with the correct state and initializing new tracks as well as detecting and ejecting spurious measurements. When multiple observations are jointly processed, a more robust data association can be obtained. This intrinsically considers the geometric relationship between a set of landmarks (see also [27]). This approach allows having a detailed representation of the environment which allows robust performance of tasks as well as efficient localization and path planning.

Last but not least, the described work in [28] presents a SLAM algorithm to automatically generate hierarchical indoor maps based on omnidirectional images. This again allows a robust performance of complexities as well as making efficient path planning possible.

Our approach is based on a particle distribution logic which is used by matching the scans

3.1 Particle Distribution Logic – General Idea

We presented the first version of our scan matcher in [29, 30]. It uses an intelligent seeding of particles in different areas. This seeding is a two-step procedure. In the first step, particles are roughly seeded in different search areas on the map with adjustable percentages of particle distribution between them. After the first seeding, the best-rated 20% of particles are selected and clustered with a k-means clustering. Within this cluster, particles are seeded again in the second step. The best-rated particle is then selected as matching pose.

To avoid jumps in the localization, a relocalization at another pose has to be stable for several iterations. As further optimization measure, the pose angles of the particle are limited to be close to the movement prediction angle.

3.2 Particle Distribution Areas

Movement Model/CAN-Based. The main particle distribution area is a quite small, circle-ring-sector-shaped area. The opening angle and the thickness are based on the movement data given by the CAN. The bigger the movement changes are, the bigger the area gets. The area is then placed where an extrapolation of the last matched pose based on the CAN data would expect the car now.

Gaussian-Based. The second area of particle distributions is meant for error corrections of the main CAN-based area. It is placed at the same position as the CAN-based area, but this area is much larger and the particles are seeded according to a Gaussian distribution. This area is used as backup, in case the CAN-based movement prediction would lead to wrong results. As it is much bigger than the main area, it should contain a lot more particles.

GPS-Based. This area is again a larger, Gaussian-distributed search area. It is localized around the position where the GPS antenna localizes the car. This area is used for global relocalization because the first two areas are only relative to the car position and accumulate an error over time. So, this area will seed particles in a known map on an approximate GPS position to globally correct the car position.

Relocalization Area. If the best-rated pose is not within the main/CAN area, a new search area, called relocalization area, is created. This area is of the same shape as the CAN area and will get the same amount of particles as the CAN area, transferred out of the particle amount for the Gaussian-based area. It is seeded around the position of the best-rated particle, which is not part of the CAN-based area.

Cluster-Based. These areas are created in the second phase of seeding, where particles are seeded again. They are located at the position of the k-means clustered best-rated particles from the first seeding. These are again Gaussian distributed search areas, but of a much smaller size than the Gaussian-based or the GPS-based search area.

3.3 Jump Suppression

To avoid a jumping of the best matching pose between nearly equal-rated poses, each relocalization has to proof itself over time. Therefore, if a best-rated particle is outside the CAN-based search area, the relocalization area is created. If the best-rated particles of the next iterations stay inside the relocalization area, the best-rated particle will be taken as best matching pose. The car will be relocated and the CAN-based search area will use this pose as extrapolation origin and the relocalization area will be discarded. As long as the relocalization area is active but not confirmed yet, the best-rated particle out of the CAN-based search area will be taken as best matching pose.

4 Cloud-Based Technology

The collection of LEMs from a fleet of vehicles in the cloud is the base for the construction of a cloud-based environment model [29, 30]. We obtain a wider range compared to the environment model based on the ego sensors of a single vehicle. All LEMs from all vehicles and journeys are combined to gain better accuracy and to correct errors.

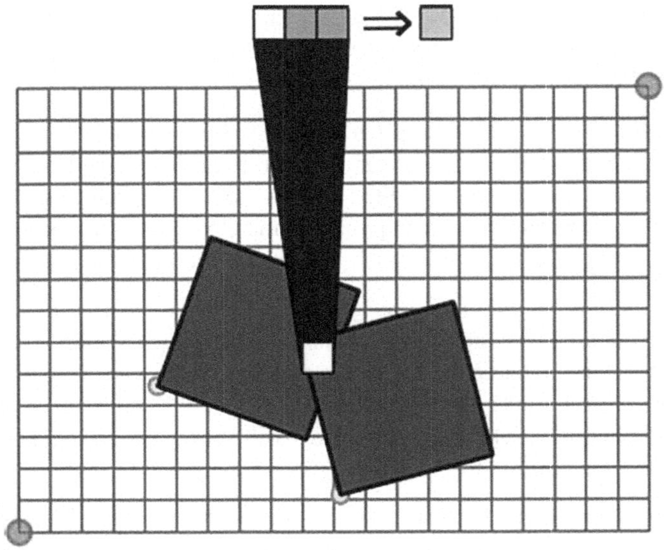

Fig. 1. Illustration of how to validate the quality of local environment models.

Merging such an enriched environment model with the local grid maps in a vehicle as illustrated in Fig. 1 allows the vehicle to use a wide range of merged sensor information in a HAD context. With this, various improvements in positioning can be achieved to enable global positioning and contents of a grid-fused map can be improved. By leveraging the cloud-based environment model, the vehicle has a more reliable environment model with global knowledge, enhanced sensor information and an ego position with an increased confidence.

4.1 Utilizing the Global Grid Layer

Producing a global grid layer out of the incoming LEM grid blocks needs to first initialize an empty graph data structure. Having acquired a suitable LEM grid block, this grid data needs to be converted into the domain model implementation. The current grid object can be added to the graph as a node to initialize the graph-based SLAM algorithm.

4.2 Graph-Based SLAM Algorithm in the Cloud

One known problem in standard SLAM is to provide a loop closure when the vehicle is crossing the same location multiple times in order to improve the quality of previous measures. Our special interest in cloud processing is not the pose of the vehicle which is delivered from a vehicle anyway via the local occupancy grid (LEM) but the pose of this local occupancy grid to obtain the position within the world. Besides that, we need to derive the relative position between two poses from the information we got out of the occupancy grid. This can be done, for example, with GPS data.

In order to match the grids, the neighboring candidate nodes are necessary to add corresponding edges to the graph. The graph then is available for a suitable output module. The final global grid layer is the superposition of all LEM grid blocks, overlapping or not, taking into consideration the corrections calculated by the graph-based SLAM algorithm.

4.3 Visualization as a Service

As new LEM grid blocks are received, the graph processor will update its current instance of the global grid layer and the graph viewer must display these changes "in real time".

Our solution is a web service that provides the visualization and controls the grid processor. Rendering of the most recent graph is performed on a central server. From here, the result is transferred to the vehicles needing environmental information. Within the vehicle, the live changing grid is displayed. A split screen offers both an overview grid as well as a detailed grid map view. The human supervisor (i.e., the driver having master rights) can zoom in, display live and playback mode of the grid processing as well as have additional information displayed.

5 Results

With the algorithms described in the previous chapters we were able to create a LEM in the vehicle based on LiDAR data. Loops are detected and handled. Figure 2 illustrates the representation of the SLAM in the vehicle.

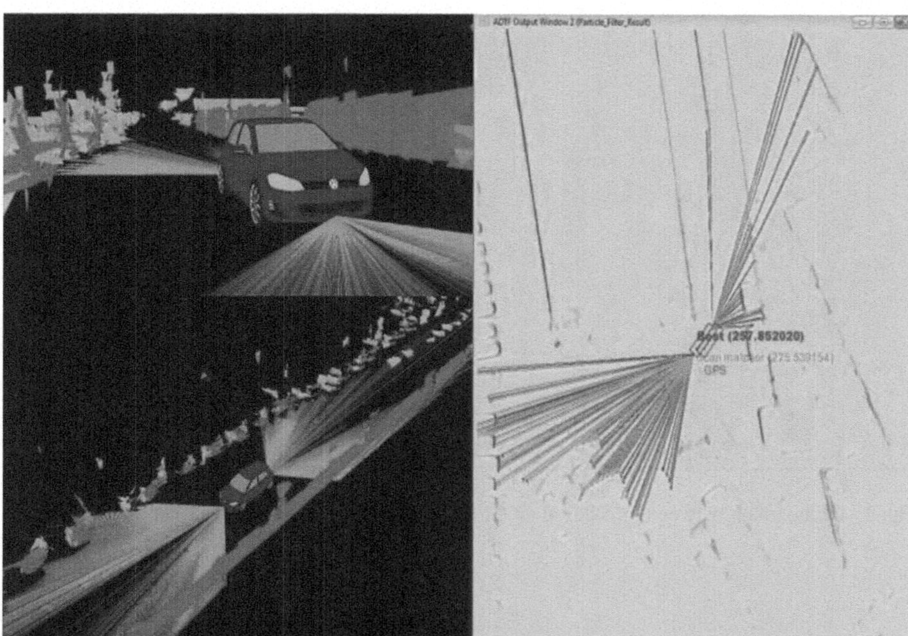

Fig. 2. Representation of SLAM in the vehicle

LEM data is sent to the cloud. In the cloud the LEMs are aggregated. A cloud-based environment model grid is created and can be visualized using a web-based viewer.

5.1 Increase Confidence of Position

Our approach is currently successfully used to increase confidence of position in a HD positioning context. Being able to give precise positioning even without GPS, this allows us to provide good indoor environment models, e.g., for automated valet parking in a parking garage.

Figures 3 and 4 show the possibility to correct an accumulated ego position error and relocalize in a known grid without GPS.

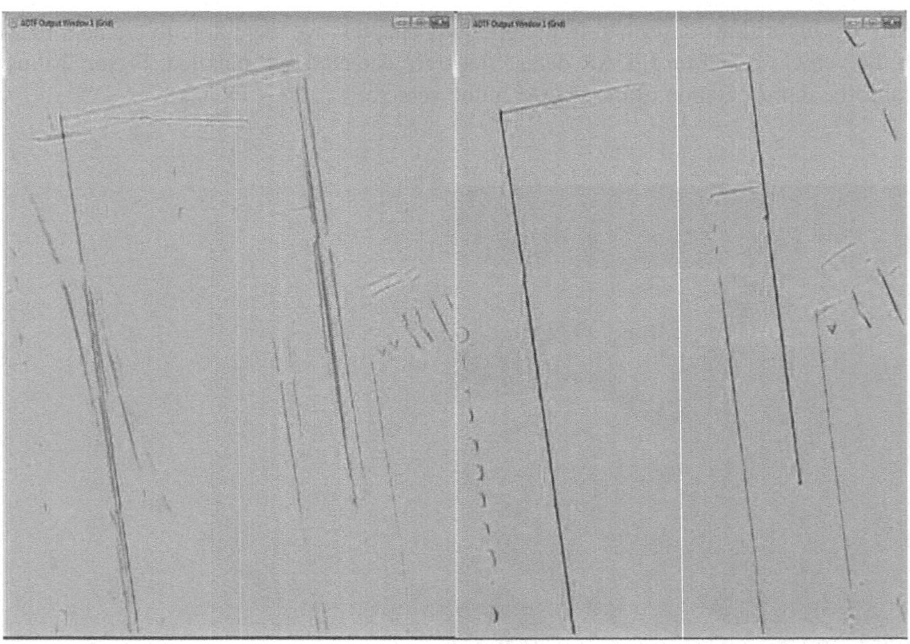

Fig. 3. Comparison between grid-based on GPS only (left) vs. grid-based on scan matcher (right)

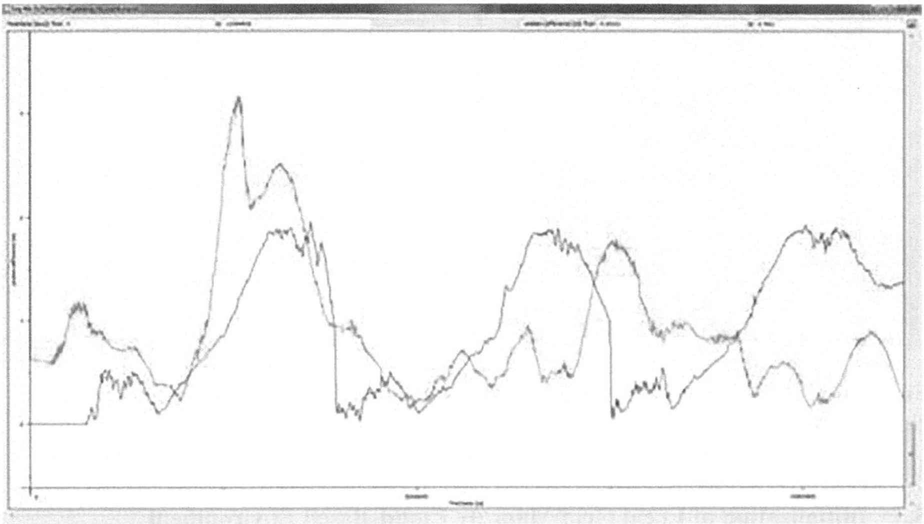

Fig. 4. Comparison between grid-based on GPS only (red) vs. grid-based on scan matcher (black) (Color figure online)

5.2 Global Correction of Local Errors

Figure 5 showcases the result of correcting local errors by cloud-based environmental information. This means that by leveraging the environment model in the cloud, the vehicle has a more reliable environment model with global knowledge, enhanced sensor information, and an ego position with an increased confidence (Fig. 6).

Fig. 5. Correction of local errors (left) by cloud-based environmental information (right)

Fig. 6. Analysis of environment models supported by visual representation of the environment models in the cloud.

5.3 Initialization of Local Grid Maps by Cloud-Based Environment Model Tiles

With the initialization of local grid maps in-vehicle by the cloud-based environment model tiles, it becomes possible to predict future areas of movement based on the currently observed movement. It also allows to download larger packages of grid blocks from the cloud and to save them. Additional benefits are to cache meta data of all downloaded blocks as well as to use downloaded data for the pre-initialization of blocks to be reached by the car as illustrated in Fig. 7.

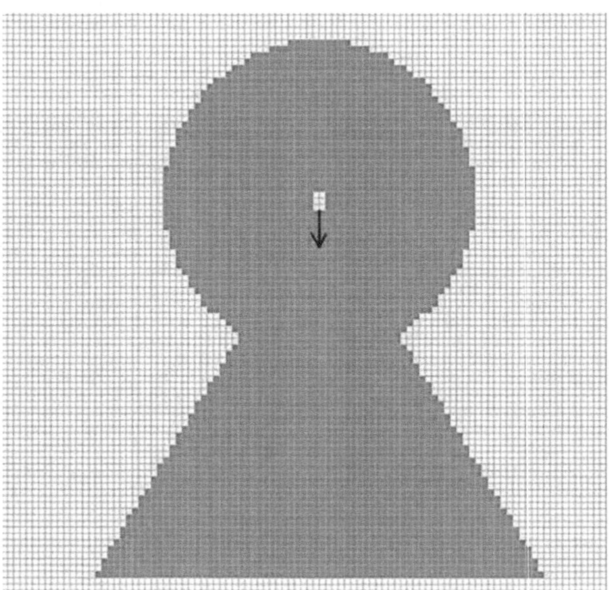

Fig. 7. Pre-initialization of future blocks for the vehicle

The client downloads the cloud-based grid packages in protobuf format. A grid package consists of 10 × 10 grid blocks with 60 m × 60 m each. The coordinate system is Universal Transverse Mercator (UTM). Only packages of one UTM zone can be combined.

A cell contains the cell position with the block as well as the normal vector. Only occupied cells are stored.

5.4 Scalable and Secure Web Service

Based on Amazon Web Services (AWS) our approach guarantees short processing times due to a parallel and stream processing. The adaptation of loads is done automatically. And last but not least the web service transfers the cloud-based environment model results to vehicles needing environmental information. Figure 8 shows the scalable backend architecture.

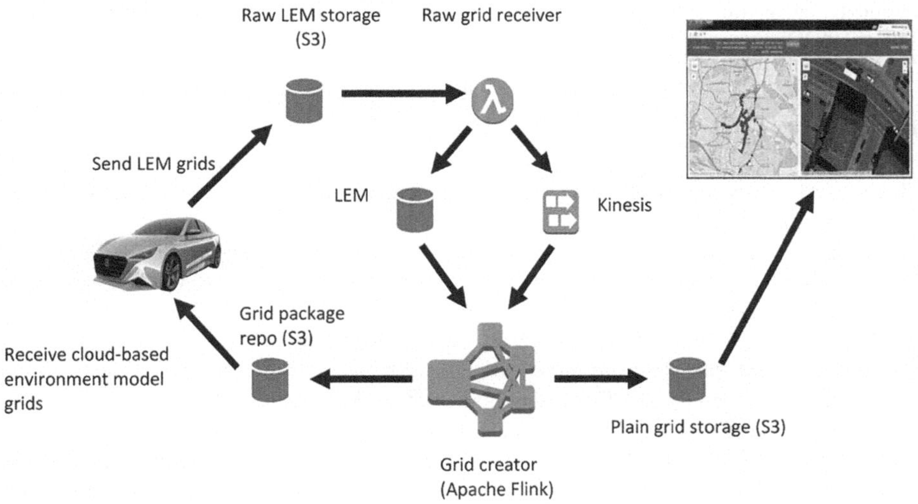

Fig. 8. Scalable backend architecture

5.5 Benefits for Data Analysts and Developers

Apart from measurement improvements, our approach also provides improvements in usability to analyse algorithms based on real data. Data analysts as well as function developers can easier investigate specific situations in order to improve their service. The visualization in Fig. 6 shows that for a developer or analyst, error handling becomes more usable with visual support.

6 Conclusion and Outlook

An environmental knowledge, good path planning, and an accurate ego position are crucial for HAD. Current approaches need to overcome limitations on GPS and interoceptive sensors, in terms of accuracy and reliability.

We have shown that our environment model approach extends the sensor range as well as the sensor availability. It allows path planning over the sensor horizon, provides additional environmental information, and improves HD positioning.

It does not need to be such a dramatic scenario as de-scribed in the introduction section: Imagine being able to leave your car in front of the entrance of your company and it will simply search and drive to the next available parking area by itself. This is made possible thanks to the shared sensor information of your colleagues' vehicles which arrived earlier; no additional infrastructure needs to be in place.

Acknowledgements. Special thanks to Holger Dormann, Olav Koska and to our project team who are responsible for the implementation of the approach presented in this paper.

Funding. Their work is partly funded by the BMVI Project PROVIDENTIA (FKZ: 16AVF1002D).

References

1. Holve, R., Dormann, H.: Big data analytics – Methoden und automobilgerechte Anpassungen. ATZ Elektronik (2016)
2. Jesorsky, O.: Kartendaten – Mehrwert für Assistenzsysteme. Hanser Automotive (2011)
3. Schneider, T.: Achieving cloud scalability with microservices and DevOps for connected cars. EE Times (2016)
4. Beringer, N.: EB Dirigo. State of the Map, Edinburgh, October 2015
5. Beringer, N.: OSM auf Rädern. AGIT, Salzburg (2015)
6. Beringer, N.: Mapping with EB Dirigo. In: FOSSGIS, Münster (2015)
7. Beringer, N.: OSM on wheels. State of the Map, Buenos Aires, November 2014
8. Beringer, N.: Das vernetzte Automobil Grenzen der Sicherheit. ATZ automobiltechnische Zeitschrift, 10 „Entwicklungen für die Fahrzeugsicherheit". Springer Vieweg, October 2013
9. Dormann, H.: Map updates for autonomous driving – the sensor cloud approach. Automotive World Webinar (2015)
10. Labella, T., Beringer, N.: Sensor based learning – one step closer to automated driving (2016). automotivemegatrends.com
11. De Rooij, F., Tremeer, B.: Enabling cars to see beyond their sensors. Automotive World Webinar (2015)
12. Beringer, N.: Sensor-basiertes Lernen: dem autonomen Fahren einen Schritt näher. In: ATZ live (2017)
13. Beringer, N.: Sensor-based learning algorithms pave the way towards autonomous driving. In: AME (2017)
14. Ludwig, J.: Elektronischer Horizont: Vorausschauende Sicherheitssysteme und deren Anbindung an Navigationssysteme. ATZ Elektronik (2012)
15. Ludwig, J.: Electronic horizon – efficiency, comfort and safety with map data. ATZ Elektronik (2014)

16. Ludwig, J.: Electronic horizon – flexible implementation of predictive driver assistance features. Technical paper (2015)
17. Holve, R.: Car as a sensor – cooperative perception and learning for automated driving. In: ATZ Conference (2015)
18. Dubitzky, R.: Das Fahrzeug als Sensor – Kooperatives Wahrnehmen und Lernen für automatisiertes Fahren. In: ATZ Elektronik, February 2015
19. El-ektrobit Automotive: Open robinos specification. El-ektrobit Automotive, Erlangen (2017)
20. Ohl, S.: From research to mass production: using a versatile platform for developing new HAD software functions. In: ATZ live (2017)
21. Boyd, R.R.: A discourse on winning and losing. Air University Library, Maxwell, AL (1976)
22. Neumann, L., Vanholme, B., Gressmann, M., Bachmann, A., Kählke, L., Schüle, F.: Freiraumerkennung Ein Grundstein der autonomen Fahrens. In: ELIV (2016)
23. Elfes, A: Using occupancy grids for mobile robot perception and navigation. IEEE Computer **22**, 46–57 (1989)
24. Moravec, M.P.: Sensor fusion in certainty grids for mobile robots. AI Magazine **2**, 61–77 (1988)
25. Castellanos, J.A., Neira, J., Tardos, J.D.: Limits to the consistency of the EKF-based SLAM. In: Ribeiro, M.I., Santos-Victor, J. (eds.) Intelligent Autonomous Vehicles (IAV-2004), IFAC/EURON. IFAC/Elsevier, Lisboa, July 2004
26. Guivant, J.E., Nebot, E.M., Nieto, J., Masson, F.: Navigation and mapping in large unstructured environments. IJRR **23**, 4 (2004)
27. Bailey, I.: Mobile robot localisation and mapping in extensive outdoor environments. Ph.D. thesis. ACFR, University of Sydney, Australia, August 2002
28. Zivkovic, Z., Bakker, B., Kröse, B.: Hierarchical map building using visual landmarks and geometric constraints. In: International Conference on Intelligent Robotics and Systems IEEE/JRS, Edmundton, Canada, pp. 7–13, August 2005
29. Beringer, N., Koska O.: Extending the sensor range for highly automated driving: a new approach for environment modelling. In: ELIV (2017)
30. Koska O., Dormann H.: Hochautomatisiertes Fahren: Die Welt als Umgebungsmodell. In: 8. Tagung Fahrerassistenz Einführung hochautomatisiertes Fahren, Munich (2017)

Efficient Sensor Development Using Raw Signal Interfaces

Martin Herrmann[1(✉)] and Helmut Schön[2]

[1] IPG Automotive GmbH, Karlsruhe, Germany
martin.herrmann@ipg-automotive.com
[2] Robert Bosch GmbH, Stuttgart, Germany

1 Introduction

A sensor's ability to accurately recognize the environment in situations of any kind is an essential requirement for the effectiveness of advanced driver assistance systems. Changing environmental conditions must never cause malfunctions, nor must complex traffic situations in a variety of surroundings. In addition to the development of the actual driving function, the recognition of the environment by individual sensors and the fusion of data from several sensors in a central environment model need to be considered. The vehicle should capture and process the environment as error-free as possible as a prerequisite for ensuring the perfect functionality of advanced driver assistance systems and automated driving functions (Fig. 1).

Fig. 1. Sensors must recognize complex traffic situations as accurately as possible

T. Bertram (Hrsg.): *Fahrerassistenzsysteme 2018,* Proceedings, S. 30–39, 2019.
https://doi.org/10.1007/978-3-658-23751-6_3

Due to the growing complexity of systems, the development of these elements requires efficient test methods. Simulation increasingly proves to be a fundamental part of the product development process since it reduces the need for real prototypes and real-world test drives during vehicle development and enables fully reproducible and automated testing of the virtual prototype in extensive test catalogs. At this point, a range of modeling depths for sensor models is necessary in order to meet the requirements of different users of sensor models.

2 Classification of Sensor Models

For the validation of automated driving functions, different functionalities such as perception, sensor data fusion as well as maneuver planning and execution need to be examined [1]. Depending on the respective functionality, various sensor model classes for the supply of environmental information are required in the simulation, while the actuator design for the vehicle control does not require sensor information but must only execute planned interventions. This results in the following tasks which the simulation should support with suitable sensor models:

- Sensor data fusion development: requires an object list of several sensors (based on different sensor technologies) that are as realistic as possible as well as ideal environmental information as a reference during testing
- Driving function development: requires an object list (ideal and/or as realistic as possible)
- Component development (radar, lidar, camera and ultrasonic sensors): requires raw data for the development of raw data processing as well as ideal environmental information as a reference during testing [2]

Basic and partly contrary requirements for a simulation environment for sensor models follow from these tasks. Large test catalogs demand a high performance and a low effort involved in the generation of the scenarios. On the other hand, it is important that raw data which is as realistic as possible and object lists are provided as sensor output – depending on the sensor technology. Due to technological advancements, it has become possible to efficiently calculate complex physical models in simulation, while in the past only very simple, ideal sensor models had to be used almost exclusively in order to ensure the real-time capability of the simulation environment.

However, as evident from the aforementioned tasks in the development of advanced driver assistance systems and automated driving functions, these complex models do not replace the simpler models based on object lists as input and output parameters. A simulation environment that may be used for all tasks thus needs to provide the following model classes.

2.1 Ideal Sensor Models

Ideal sensor models represent a generic interface for the extraction of information from the environmental simulation. Irrespective of the technologies used in the real world (e.g. camera, radar, lidar, ultrasound), object information is provided within the acquisition range of the sensor. Thus, acquisition of the environment is ideal without any detailed consideration of physical, technology-dependent effects.

Models like these are applied in early stages of function development and in rapid prototyping when the technology to be used has not yet been determined or is to be disregarded. Furthermore, errors on the perception level can be excluded during the test of the function as well and thus errors of the driver assistance or driving function be examined in detail. In addition, an ideal sensor model can serve as a reference for signal processing algorithms based on raw data for sensor fusion applications [2] (Fig. 2).

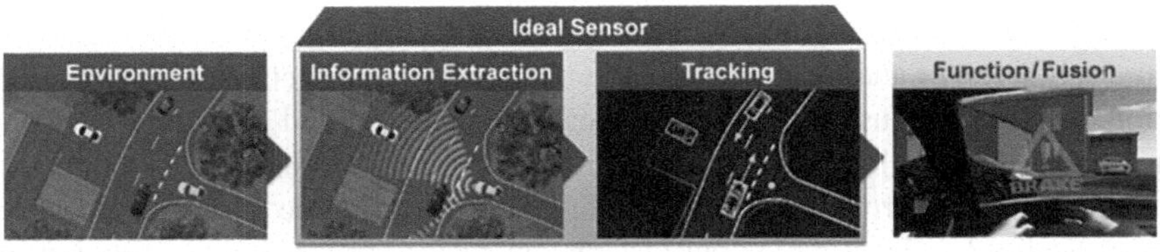

Fig. 2. Ideal sensor for rapid prototyping/function development (blue: simulation tool, yellow: user) (Color figure online)

2.2 Phenomenological Sensor Models

Phenomenological sensor models deliver a higher level of detailing. They use ideal environmental information and superimpose technology-dependent effects known from theory and measurements onto these.

The modeled effects have a negative impact on the ideal object list. This may be caused by false negatives (lacking detection of existing objects), false positives (detection of non-existent objects, i.e. "ghost objects") as well as a manipulation of the virtual measurement data (e.g. stochastic noise, determination of the position/speed within the measurement accuracy) [2].

Ideally, a phenomenological sensor model is designed in a way that allows it to be parameterized using data sheets or simple measurements of the real sensors to be simulated. No detailed system knowledge is required to be able to use a phenomenological sensor model. Typically, it is used for the development of sensor data fusion algorithms, driver assistance functions as well as automated driving functions.

The resulting object list contains signals that correspond to those which the imitated technology would make available in reality after processing the sensor-internal raw data without taking implementation artifacts into account. In order to do so, the environment model requires additional parameters which influence the impact of the modeled effects. For example, detectable objects may be equipped with the pre-calculated or measured angle-dependent property of radar backscattering (RCS map) for the modeling of radar reflections.

In comparison to ideal sensor models, however, the additional calculation effort caused by these effects is much lower than in the case of physical sensor models on the raw data level. This sensor model class thus enables efficient testing of the function including any sensor-specific effects as well as the development and testing of data fusion of several different sensors (Fig. 3).

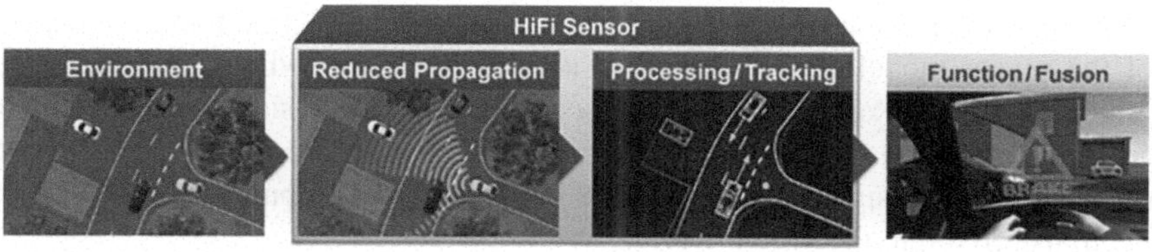

Fig. 3. HiFi Sensor for function development and testing (blue: simulation tool, yellow: user) (Color figure online)

2.3 Physical Modelling of Signal Propagation Using a Raw Signal Interface

Physical models with a raw signal interface take into account the actual signal propagation including the most relevant physical effects in the interaction of the signal with objects and the transmission media along the propagation path. The modeling of wave propagation throughout the environment is called a "channel model" in the case of a radar sensor, for instance. Channel modeling aims at making raw signals available in a form in which they are present at the interface between the receiving device and the component control unit in reality.

Thus, depending on the technology that is modeled, realistic image data, ultrasound echoes or channel impulse responses, for instance, are generated for the radar in the simulation, aside from implementation artifacts of the sensor [3].

This allows for an early testing of raw data processing which is executed in the ECU on the sensor in reality. Component developers implement corresponding models in MIL/SIL environments or complete this processing on the real control unit on a HIL test bench. Only this second module completes the sensor model and provides the object list required for the function. At this point, profound system knowledge is necessary [4].

Due to the detailed modeling of signal propagation, these models are generally significantly more complex than ideal or phenomenological sensor models. Only with this class of sensor models can the complete event chain of sensor signal processing and the function to be developed be investigated. The environment model needs to exhibit an appropriate level of detailing in order to supply the physical effects taken into account in the channel model with adequate input.

This leads to a considerable effort necessary for the modeling of the scenario. Here, the quality of the 3-D models, the level of detail and the wealth of variants as well as the material parameters that are assigned to the individual surfaces of the 3-D models are imperative (Fig. 4).

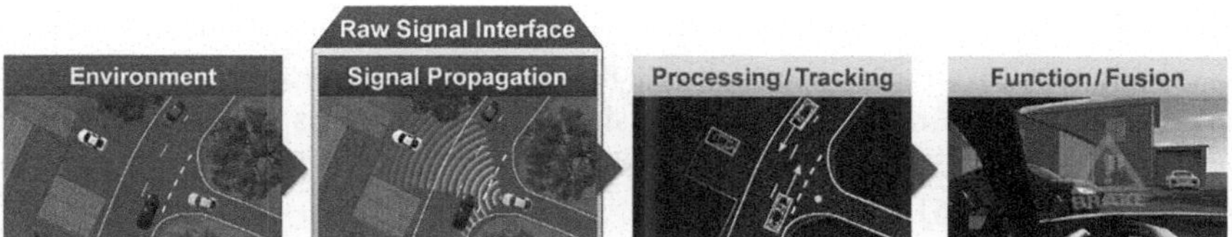

Fig. 4. RSI for component development and testing (blue: simulation tool, yellow: user) (Color figure online)

The open integration and test platform CarMaker by IPG Automotive supports all of the sensor model classes described above. In the following section, the radar raw signal interface (Radar RSI) is addressed in more detail. Using a realistic channel model with a raw signal interface in combination with a detailed sensor model for the sensor hardware and signal processing, the component development of radar sensors can be transferred into the virtual world.

The optimization of the mounting position, the antenna design and the signal processing test, which is executed within the ECU of the physical sensor in reality, thus become possible at a much earlier time in the development process. For a better understanding of the model requirements, the general functional principle of a radar sensor is first considered.

3 Functional Principle of a Radar Sensor

Radar sensors transmit electromagnetic waves and subsequently receive the backscattered signal (impulse response). The angle-dependent reception performance depends on the antenna properties, the path length, and the backscattering properties of the scattering objects. The amplification of the receiving antenna, which does not necessarily have to be the transmission antenna at the same time, has an influence on the received signal strength [5].

When a signal is transmitted by the radar sensor, the signal must be marked so that it can be clearly identified upon its return and its signal transit time be analyzed for example. This identification possibility is usually provided by the transmitter via amplitude or frequency modulation and demodulated again when the signal is received. Following the demodulation, findings about the distance of the detected object (based on the signal transit time) and about the relative radial velocity (based on the frequency shift due to the Doppler effect) can be obtained [5].

The detection of the horizontal position (azimuth angle) of objects may be achieved via slewing of the radar beam, several transmission and receiving antennae, or several radar sensors on the vehicle which are distributed across the vehicle's width. The reception and digitization of the analog signals is followed by signal processing. Via a clustering of individual reflection points of the detected objects, a so-called object hypothesis can be made. This is followed by a continual tracking of the assumed object. The detection quality is enhanced by means of the time curve of the measurements [5].

There are other effects that may have an influence on radar measurements. Due to so-called multipath propagation, reflections on guardrails, lateral panels (especially on trucks), road surfaces, etc. lead to incorrect distance measurements or even ghost objects (false positives). Different objects, depending on geometry, material and relative orientation toward the sensor, have different radar cross-sections (reflectivity). In terms of maximum range, the transmission power decreases as the inverse fourth power over the distance and additionally is negatively affected by atmospheric effects such as precipitation [5].

Particularly relevant for automotive applications is the simultaneous detection of several objects (multi-target capability). This requires a high resolution capability in the dimensions of distance, azimuth angle, and relative velocity. If several objects do not sufficiently differ in any of these dimensions, this may lead to the radar sensor erroneously perceiving them as a single object (merging) and thus classifying them incorrectly [5].

4 Radar RSI

The development of radar sensors includes the sensor hardware (receiving and transmission antennae and ECUs) as well as the sensor software (raw signal processing, object hypothesis, classification and tracking as well as driving function).

Signal propagation beyond the sensor (propagation of the electromagnetic wave and reflections on surrounding objects) does not depend on the sensor's design and is subject to the general laws of physics. It is thus useful to make a channel model for this signal propagation available in the simulation environment. In simulation, this is achieved by discretizing the wavefront into a beam and calculating the reflections and propagation losses.

For the sensor model, multipath propagation and double reflections (repeated path propagation), for example, need to be taken into account (see Fig. 5). These effects result in misleading information about the angle or distance which sensor signal processing aims to correct by means of complex object tracking among other things. The need to model multiple reflections in signal propagation results from these effects.

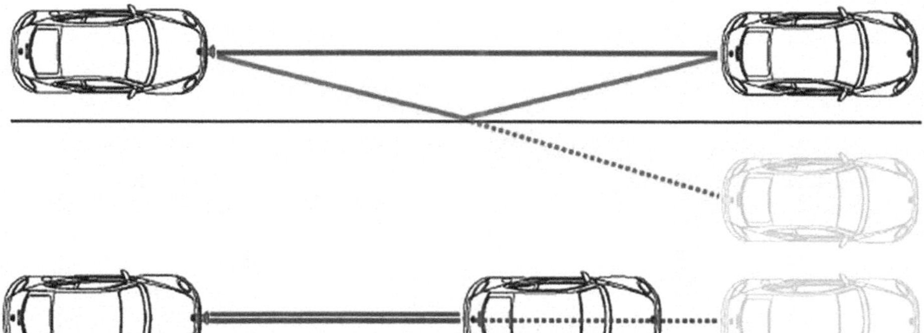

Fig. 5. Multipath propagation on a guardrail (top) and repeated path propagation (bottom)

For the calculation of the reflections, the material properties as well as the complex geometric shapes of the static and dynamic 3-D objects in the scene and their kinematic information are required. The realistic modeling of the scene alone considerably increases the parameterization effort required for virtual tests in comparison to the use of the other two sensor model classes. It must be ensured that all 3-D objects in use have been accurately modeled and prepared for channel modeling. This applies to fully closed surfaces, correct normal vectors on the surfaces as well as radar-specific material properties (in this case permittivity and scattering behavior of the material). For example, the permittivity of steel is very high, while the scattering of radar signals on metallic surfaces is very low; asphalt, in contrast, exhibits very low permittivity and pronounced scattering properties.

In wave propagation, the Doppler effect occurs when a relative radial velocity exists between the transmission antenna and a reflecting object. The frequency shift resulting from all reflections along the propagation path may be used by signal processing in order to determine the speed of target objects using the speed of one's own vehicle. Here, the fact that only the radial velocity component can be determined needs to be taken into account.

Propagation losses are factored into the simulation along the total distance traveled while effects caused by the atmosphere, rain or fog are modeled. For reflections on surfaces, losses are also modeled which depend on the relative permittivity of the material as well as the polarization and the direction of the incident wave.

Ultimately, the channel model provides the channel impulse response for further processing. This information must subsequently be analyzed by the user while considering the antenna properties in order to implement the user's own processing and tracking – even before a sensor is physically available (see Fig. 6). Correspondingly, the user is also required to have very detailed knowledge of the functional principle of the radar sensor to be modeled for the implementation of the processing chain. It is also possible to exchange models as black boxes – e.g. as a Functional Mock-up Unit (FMU) – between the developers of the OEMs and their suppliers. This enables function developers to have access to a complete sensor model while the intellectual property of the component developer remains protected.

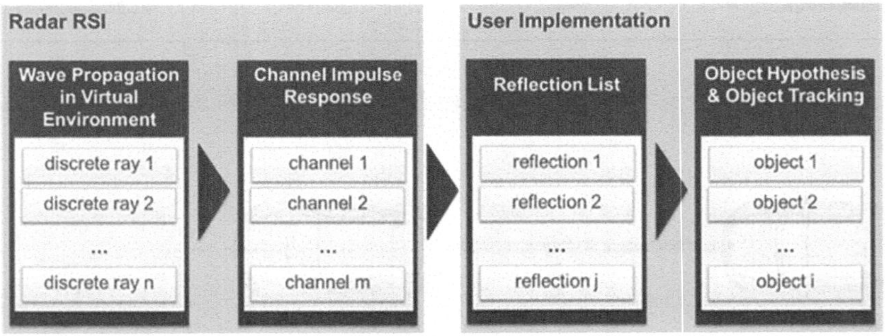

Fig. 6. Schematic overview of the use of the Radar RSI and the user's tasks

5 Application Example

In order to model the full event chain, a reflection list is first generated from the information transmitted by the raw signal interface, which then has to be converted to an object list through object hypothesis and a tracking module.

The application of a radar raw signal interface is demonstrated here by means of a simple scenario. In this example scenario, an ego vehicle drives on the right lane of a three-lane highway while approaching a slower target object. Meanwhile, a faster vehicle driving on the center lane overtakes the ego vehicle. Figure 7 shows the scanning of the scene by two antennae, blue indicating close range and red long range.

Fig. 7. Screenshot from an example scenario in CarMaker (Color figure online)

In this example scenario, the raw signal generated by the Radar RSI module was used as input for a physical sensor model by Bosch from which a reflection list of the example scenario was created. This requires taking into account the antenna properties as well as modeling the conversion of the analog signal into a digital signal by simulating the hardware involved.

In addition, digital signal processing is necessary which – if possible – directly models the algorithm used in the sensor. Each radar reflection contains information about the distance in longitudinal direction (x), the azimuth angle or an estimation of the distance in lateral direction (y), and the relative radial velocity v_{rel} to the detected object.

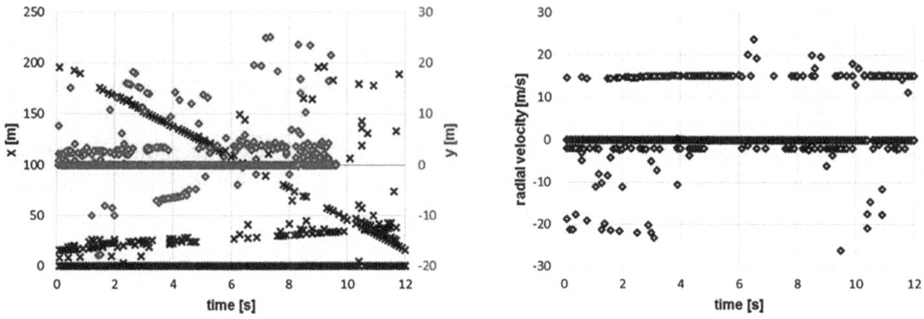

Fig. 8. Graph of the reflection list (x(t), y(t), v_{rel}(t)) from the radar sensor's signal processing over time

The graphs in Fig. 8 show a clear illustration of the resulting reflection list. In the graph which illustrates the distance to the target objects, the increasing distance in the direction of x to the overtaking vehicle is evident. The detection begins at a distance of approximately 10 m once the target object enters the radar lobe. Furthermore, it is evident that the object directly ahead of the ego vehicle is reliably recognized from a distance of approximately 170 m. In some cases, distances slightly longer than the actual distances are estimated. This is a result of multiple reflections via the adjacent guardrails. Other reflections at very close range of the ego vehicle are clutter caused by scattering on the road surface.

The lateral offset of the vehicles is also shown in this graph. The vehicle ahead is correctly detected with only minimal transverse offset; the overtaking vehicle with an increasing transverse offset of approximately 2.5 to 5 m. Incorrect estimations of the transverse offset caused by multiple reflections via the guardrails are evident.

The second graph shows that the relative velocity between the ego vehicle and the overtaking vehicle is very consistently estimated at approximately 2 m/s. The vehicle ahead is approximately 15 m/s slower than the ego vehicle. This information is also provided very reliably.

At the present time, the generation of the reflection list has been implemented. The object hypothesis and tracking will be added in the future in order to model the full sensor model in the simulation as well and demonstrate the closed-loop integration of a driving function.

6 Outlook

In recent years, the test effort in vehicle development has continually risen due to the growing diversity of vehicle variants. Particularly in the field of advanced driver assistance systems and automated driving, there are countless test cases. Here, changing environmental conditions must never cause malfunctions, nor must complex traffic situations on various route sections. In order to be able to reliably evaluate the operability of the systems, tests for any conceivable traffic situation must be conducted on all types of roads. Covering all these test cases in reality is very difficult, if not impossible.

This difficulty can be overcome with efficient test methods, e.g. by means of simulation. The essential role that sensors play for automated driving functions imposes specific requirements on a simulation environment, such as a very high level of detailing of all environment models, for example. CarMaker supports three sensor model classes: ideal and phenomenological sensor models as well as a physical channel model with a raw signal interface. This presentation addressed the radar raw signal interface (Radar RSI) which transmits the channel impulse response by calculating reflections, propagation losses and frequency shifts.

This raw signal modeling approach can be adapted to other sensor technologies (e.g. lidar). In this case, different physical effects must be modeled accordingly and the signals provided by the interface must be adjusted to the technology modeled. The existing 3-D model library serves for continued use; only the material database requires further technology-specific entries. Sensor development may thus focus on modeling the sensors and optimizing signal processing.

One important aspect is the validation of the sensor model in order to verify the extent to which the model corresponds to the real sensor behavior. For this, a scene previously driven in reality must be modeled in the simulation and the measured sensor data be compared to the simulated sensor data. Not only are the quality of the sensor model but also the correct transfer of the real environment into the simulation environment as well as the realistic specification of wave propagation decisive here. Significant advancements have already been made in all of these issues as described in this presentation using the sensor models in CarMaker as examples.

However, in order to be able to release driving functions via a simulation of the environment and the sensor alone, further improvements are necessary which will continue to pose a considerable challenge in the foreseeable future.

References

1. Abdellatif, H., Schick, B.: Virtuelle Homologation von softwareintensiven, sicherheitsrelevanten Systemen: vom ESP bis zum Automatisieren Fahren. In: Internationaler Kongress ELIV, Bonn (2017)
2. Herrmann, M., Dörr, D.: Echtzeitfähige Sensormodelle für den virtuellen Fahrversuch: Klassifikation und Anwendung in Entwicklung und Absicherung automatisierter Fahrfunktionen. In: Internationaler Kongress ELIV, Bonn (2017)
3. Henning, J., Herrmann, M.: Virtual test and validation using real-time capable physical sensor models. In: Autonomous Industrial Vehicle Technology Symposium, Cologne (2017)
4. Pfeffer, R.: Advanced sensor simulation technology for autonomous driving development. In: NIWeek, Austin (2017)
5. Skolnik, M.: Radar Handbook, 2nd edn. McGraw-Hill Education (2008)

360° Surround View Radar for Driver Assistance and Automated Driving

Dietmar Stapel(✉), Carsten Roch, Helgo Dyckmanns, and Martin Mühlenberg

HELLA GmbH & Co. KGaA, Lippstadt, Germany
dietmar.stapel@hella.com

1 Evolving Market Regarding Automated Driving

Fully autonomous vehicles (AVs) are unlikely to be commercially available before 2020. Meanwhile, advanced driver assistance systems (ADAS) will play a crucial role in preparing regulators and consumers for the predictable reality of vehicles taking over the control from drivers.

Today, a large number of driver assistance systems are available for almost all vehicle segments. They ensure stability in critical situations, maintain a safe distance to the vehicle in front, and support the driver while parking. Regarding automated driving, the industry is still in the initial phase and faces a steep uphill climb in terms of market penetration referring to the latest analysts' forecasts. Despite the matter of time, the progress in driver assistance systems will accommodate the mandatory base, technologically as well as commercially, for the upcoming development towards

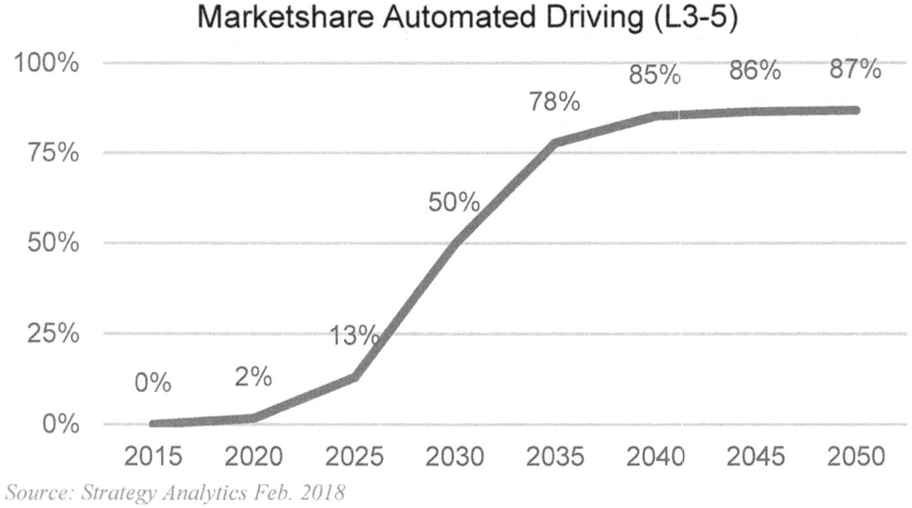

Fig. 1. Market forecast automated driving for SAE LEVEL 3–5 vehicles (© HELLA)

T. Bertram (Hrsg.): *Fahrerassistenzsysteme 2018,* Proceedings, S. 40–46, 2019.
https://doi.org/10.1007/978-3-658-23751-6_4

highly automated driving. Market analysts, as shown in Fig. 1, predict that 50% of passenger cars being sold in 2030 have the potential to be highly automated (SAE Level 3–5) and 15% of new cars sold in 2030 could be fully autonomous (SAE Level 5). This promising outlook constitutes a bright future in terms of market potential, but also implies various challenges along the way.

2 From Driver Assistance to Automated Driving

Taking the introduction of ADAS systems as lessons learned, the challenges for the market penetration of highly automated driving are system costs as well as the acceptance of the consumer. This results in an evolutionary approach towards highly automated driving. Whereby the current democratization of driver assistance functions, driven by legislation or safety programs like EuroNCAP, pressures the industry to keep the ADAS cost low and further diffuse features from premium segments into the volume market. Moreover, the progressing expansion of the functional spectrum, as well as the blending of individual ADAS functions to superordinate systems, leads to a gradual increase of the vehicle autonomy. This stepwise approach provides enough time and touch points to build up customer acceptance and consequently amplifies the customers willingness to purchase automated driving features.

The currently utilized sensor technologies for ADAS functions like ultrasonic, camera, radar and the imminent lidar sensors, usually operate independently resulting in a decentralized decision-making process of the systems. A decentralized approach is applied successfully for traditional driver assistance applications, where the driver is usually being informed about dangerous situations by a warning signal. Higher automation, however, requires a combination of the measured data where a various number of sensors deliver all relevant environmental data to a centralized system that implements the ADAS functions. This so-called sensor fusion and the resulting possibility to create an environmental model are seen as a necessary basis for the realization of highly automated driving functionalities. This, in turn, implies new requirements regarding sensor communication and architectures, since a high amount of data has to be distributed, merged, and interpreted.

By following the current market demand with the parallel presence and introduction of ADAS and highly automated driving, the sensor requirements regarding corner radar sensors, schematically displayed in Fig. 2, need to be satisfied in a modular and scalable manner. A favorable price-performance ratio respectively radar sensor performance is achieved through a number of approaches: highly integrated components for more compact and economically constructed sensors, high-performance microcontrollers with increased memory and computing resources to implement more complex algorithms, and the expansion of the available bandwidth by using a higher frequency range and innovative modulation concepts (76–81 GHz).

The resulting scalable platform ensures a cost-effective development of sensor solutions that provide extra performance for system growth and permits the flexibility to integrate new features with minimal design changes.

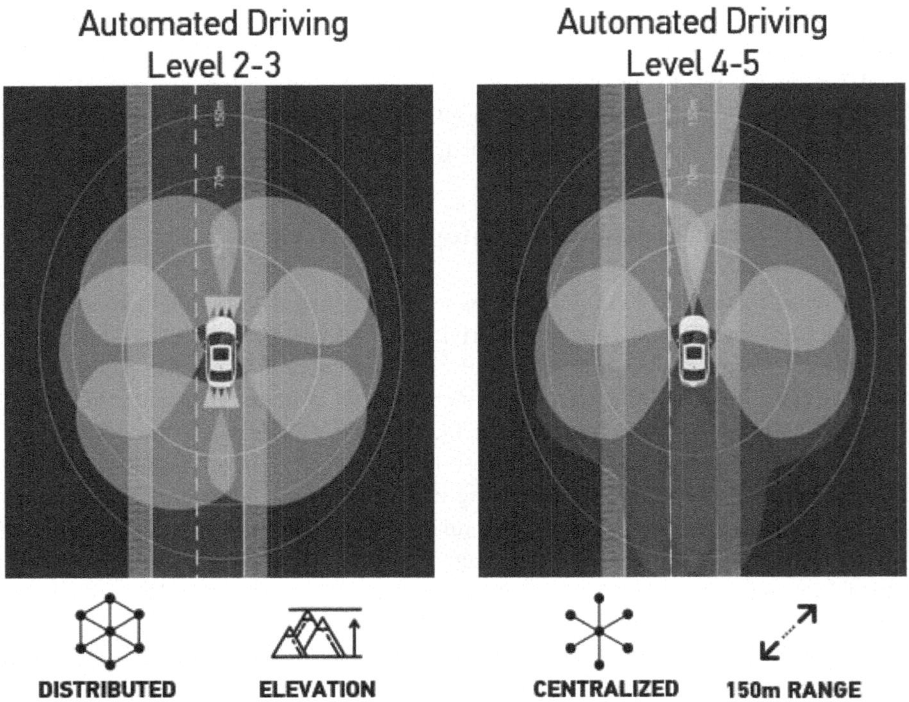

Fig. 2. Radar sensor requirements for different levels of autonomy (© HELLA)

3 Radar Sensors for Automated Driving

Radar-based sensors represent a central technology in the context of highly-automated driving, since these sensors are sufficiently robust and deliver reliable results independent to weather and light effects.

Current requirements regarding environmental perception follow the predictable trend towards a radar-based 360° surround view. Therefore, a centralization of the system architecture, compact sensor dimensions for integration in the side area, as well as a flexible configuration are key factors to fulfill these requirements.

The 77 GHz corner radar sensor family by HELLA, Fig. 3, follows with its modular and scalable platform these versatile market requirements. Due to the compact dimensions the sensor is suitable for integration into the side area of the door as well as in the rear/front corners of a vehicle and thus enables the necessary 360° field of view for autonomous driving. The compact design was realized by means of the highly integrated Radar System Chip (including RF-Rx/Tx circuit, AD converter, self-diagnosis, digital interface) and thus leads to a flexible and cost-effective sensor solution.

Fig. 3. HELLA 77 GHz radar family (© HELLA)

The RF-CMOS based Radar System Chip takes over the generation and processing of the high-frequency signals but supports, compared to the common SiGe BiCMOS solutions, a higher node density and therefore the advantage to integrate additional digital signal processing components. A further benefit of this technology is the relatively low power consumption and thus, among others, improved heat dissipation of compact designed devices.

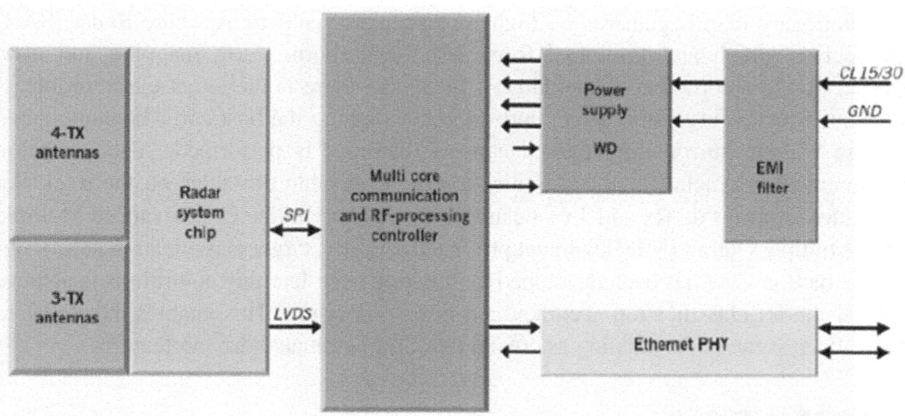

Fig. 4. Radar system architecture using a highly integrated radar system chip (© HELLA)

The sensor architecture shown in Fig. 4, illustrates an exemplary adaption of a RF-CMOS-based 77 GHz technology. The Radar Processing Controller transmits the parameters for the frequency modulation of the radar signal via a serial peripheral interface (SPI) to the Radar System Chip. Depending on the driving situation and the function, the characteristics can differ due to different requirements for range and resolution. For example, a parking function as near-field application requires a different configuration compared to an intersection assistant, in which a higher range performance is mandatory. Correspondingly, the Radar System Chip operates in different bandwidths and chirp times.

Since all radar functions are integrated in one chip, not only a compact single-circuit design made of Radar System Chip and controller is possible, but also disturbances and information losses (noise) that can occur with analog transmissions are reduced. The system for self-diagnosis also continuously monitors all functions—an important prerequisite for highly-automated driving.

4 Radar Based Environment Perception for Automated Parking

Radar sensors generate target points with relative speeds, distances and angles. Depending on the driving situation, the data is interpreted and tracked in order to anticipate the environment. This results in an environmental perception of dynamic and static objects, Fig. 5. The measured data is used to generate or further augment an image of the environment (central environment model), which serves as the necessary basis for automated driving functionalities like parking.

From a signal processing perspective, three basic measurement capabilities are mandatory for increased scene understanding by automotive radar. First, there is the stationary target mapping, which has the objective of precise mapping of the host vehicles stationary environment to address functionalities like parking space measurement, where the exact parking space dimension is derived from the stationary surroundings. This stationary map is gathered by high-sophisticated Synthetic Aperture Radar (SAR) approaches, which are developed from aerial-/spaceborne earth mapping and now, applied by the automotive radar industry. Secondly, there is the target elevation measurement, which is crucial to detect curbstone for aligning the host car, when automated parking without surrounding parked cars as reference is performed. Also, elevation measurement is crucial for the classification of traversable obstacles on the road. For this, antenna arrays on Rx and Tx side are developed and utilized via advanced Multiple Input Multiple Output (MIMO) signal processing. Lastly, target classification at low signal abstraction level has been developed by utilizing deep-learning algorithms to achieve state of the art classification accuracy and time-to-decision. This enables the Radar to gather deep scene understanding as crucial input for automated driving features.

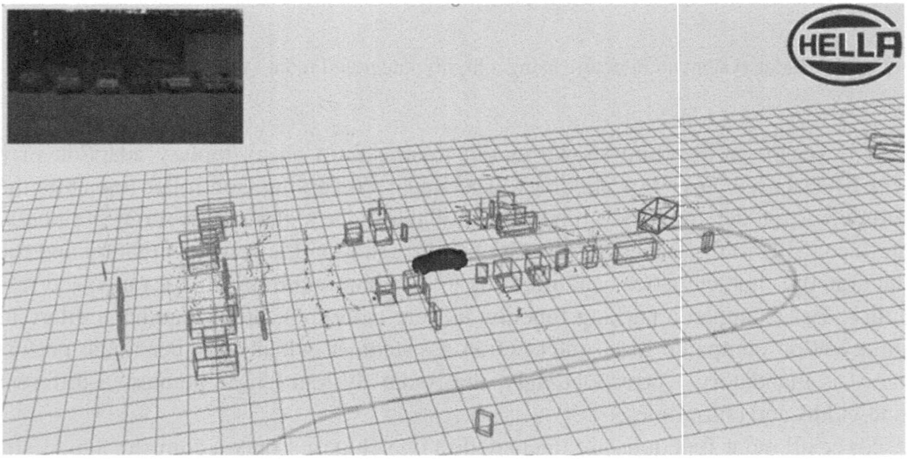

Fig. 5. Radar based environment perception (© HELLA)

5 Intelligent Validation of Automated Driving Features

With the introduction of automated driving, the automotive industry breaks new grounds not just in functionality, but also in terms of validation. Even extensive road tests cannot statistically prove the safety of an automated vehicle. This makes the use of alternative validation techniques necessary. In principle, these techniques are already known, but in order to apply them many fundamentals must still be determined, especially the base amount of required field data. Thus, methods for obtaining this data from road testing and field applications gain a high importance for future safety certification as a basis for the approval of vehicle automation systems.

Therefore, new accelerated evaluation processes need to break down difficult real-world driving situations into components that can be tested or simulated repeatedly, exposing automated vehicles to a condensed set of the most challenging driving situations.

HELLA Aglaia Testing Solutions enables this approach by initially standardizing relevant "critical" test cases and scenarios with applicable metrices and key performance indicators (KPI). High performance data acquisition, based on modular logging systems, are then utilized as input for a world-wide distributed, cloud-based, data annotation systems. The resulting distributed and standardized test data base provides scenario centered measurement data to development and test teams world-wide. These virtual scenarios accommodate the necessary basis for highly automated software in the loop (SiL) and hardware in the loop (HiL) testing and can be applied for various test executions and evaluations as indicated in Fig. 6.

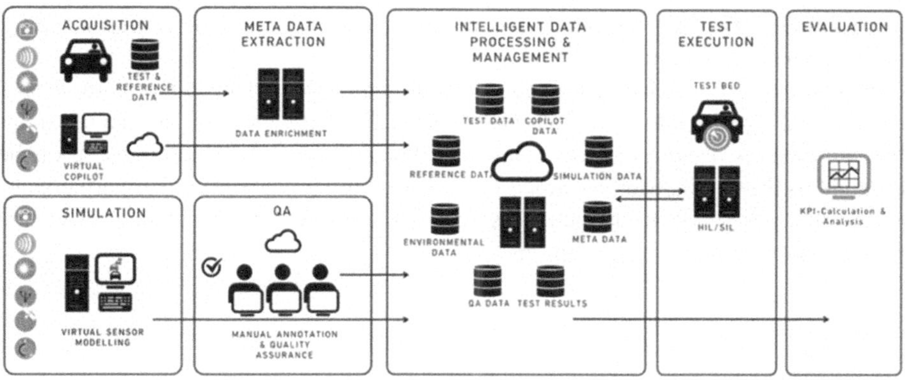

Fig. 6. Perception performance validation—process & toolchain (© HELLA)

6 Partnerships as Key to Accelerate Development

The industry is evolving from competition among individual players towards new competitive interactions in form of partnerships in open, flexible ecosystems to accelerate and reduce development efforts. This leads to OEMs, suppliers, and service providers forming partnerships across the industry and benefit jointly from sharing their individual experiences as well as the necessary infrastructure.

As the entire industry is undergoing transformational changes driven by the key trends like autonomous driving, players must adapt their organization to facilitate and implement successful collaborations. On one hand, there is the direction to emphasize partnerships on component level to transform bleeding edge technology to sensor solutions feasible for the volume market. On the other hand, there are mandatory partnerships across the industry to accomplish an efficient development in the extensive and complex framework of autonomous driving. HELLA and ZF for example entered into a strategic partnership where both automotive suppliers benefit from the cooperation in various ways. ZF, in this case, is further strengthening its portfolio as a system supplier, whereas HELLA will drive technological development and benefits from a broader market access with its leading sensor technologies. While this joint development is focused on systems solutions, each partner continues to develop and offer its technology independently. Therefore, the true potential of a long-term development partnership can be utilized in which both partners have the ability to drive common system architectures as well as product families adapting to each other.

7 Summary and Outlook

With driver assistance systems as predecessor, automated driving is in its initial stage and their future deployment is expected to accelerate gradually. The capabilities of sensors as well as data processing are continually growing and enable the monitoring of vehicles surroundings in real time to generate a gap-free and redundant environment model as basis for automated driving. HELLAs modular and scalable 77 GHz radar sensor family further drives the realization of highly-automated driving with its capabilities of a 360° surround view and the advanced near field performance for automated parking features.

In parallel, the validation framework is progressing with machine learning and artificial intelligence into a feasible and manageable fundament of safe and sustainable autonomous driving features. This contributes to the complex ecosystem of the autonomous driving industry in which players need to form collaborative partnerships to share the necessary infrastructure as basis for the development of innovative and efficient autonomous driving solutions.

Overall Approach to Standardize AD Sensor Interfaces: Simulation and Real Vehicle

Carlo van Driesten(✉) and Thomas Schaller

BMW Group, Munich, Germany
thomas.sa.schaller@bmw.de

Abstract. The standardization of subsections of complex hardware and software setups with the focus on automated driving functionality allows cost reduction in development, test and validation, as well as a simplification of component test infrastructures involving many different partners. The current VDA initiative for an ISO standard (ISO 23150) towards standardized hardware sensor interfaces [1] was mainly motivated by the struggle of dealing with large sensor setups in cars for functions enabling automation level 3 and above. The integration of Radar, LIDAR, Ultrasonic and Camera systems amongst others from various suppliers, including the component tests of up to 50 sensors for fully automated driving, demand common test procedures. These are complicated by contradictory interpretation and description of the commonly perceived environment. This paper highlights the potential benefit of this undertaking by enlarging the scope towards other test instances involving the complete chain of effects focusing on Software in the Loop (SIL) using sensor models for development, verification and validation. Therefore, the Open Simulation Interface (OSI), endorsed by the Pegasus Project [2], is introduced as the first reference implementation of the upcoming ISO standard for rapid prototyping and the common development of sensor models.

1 Motivation for Sensor Interface Standardization

1.1 Complex Sensor Setups for AD Functions

Environment perceiving sensors are the key information collecting components, besides car2x communication, to independently build the **Environment Model**; a virtual representation of the vehicles surrounding [3] on which core automated driving (AD) functions such as a highway assistant can act upon. This entails a full coverage around the car with high sensing ranges simply considering the relative velocities of cars driving towards each other, combined with the necessity to detect them early enough. In addition, the potential malfunctioning of individual sensors along with the physical constraints of each technology demands redundancy in the overall system setup. The corresponding functional safety requirements are defined by the ISO16949 with its ASIL ratings for electronics and programmable systems. The inability to detect objects with certainty by specification due to for example bad weather

T. Bertram (Hrsg.): *Fahrerassistenzsysteme 2018*, Proceedings, S. 47–55, 2019.
https://doi.org/10.1007/978-3-658-23751-6_5

conditions is otherwise not subject of this standard [4]. Optimizing overall certainty is achieved by fusing the information of each sensor, taking their individual strengths and weaknesses into account. An overview of the commonly used sensor technologies is shown in Fig. 1 and indicates the compensation achieved through sensor fusion.

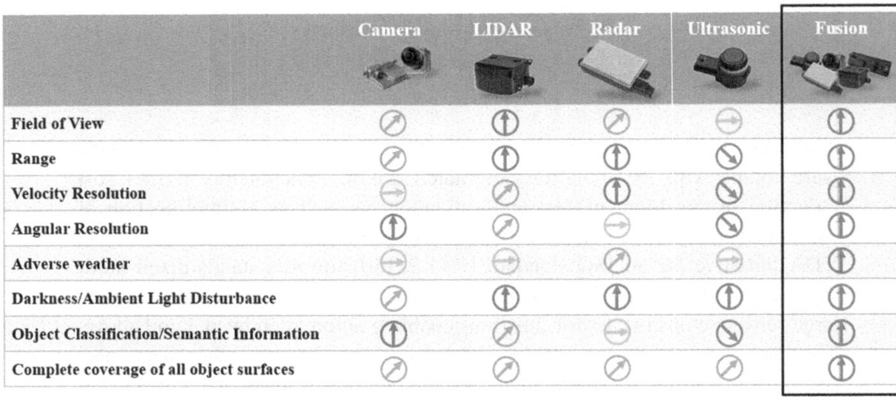

Fig. 1. Comparison of different sensor technologies [1]

A car relying on surround view cameras as its only sensor technology would render the AD function blind in a foggy environment, therefore the need to use of another complementing technology like Radar becomes apparent in this example. The integration of Radar, LIDAR, Ultrasonic and Camera systems from various suppliers demands common testing procedures, especially if the component tests of up to 50 sensors for fully automated driving are regarded. These are often complicated by contradictory interpretation and description of the sensors commonly perceived environment. Additionally, suppliers for specific sensor technologies can change over time which leads to the conclusion that common interfaces, ideally allowing plug-and-play of a sensor from another supplier or even one based on another technology, is a hard requirement for every car manufacturer.

1.2 Standardization of the Logical Interface from Sensors to the Fusion Unit

The fusion unit is part of an architecture in vehicles offering AD functions which can be described by the signal flow in Fig. 2. It shows the stepwise operations a system has to fulfill in order to trigger an arbitrary function due to an observation by sensors.

The lowest level describes a set of different sensor technologies converting (electrical) **analog signals into digitalized signals**. This is handled by the sensor itself and generates, for example, a stream of images for a Camera or a base-band spectrum in a Radar sensor. These signals are further processed, and as a first step the **Detections** are extracted. Detections can be seen as reflection peaks in the Radar sensor for certain frequencies and radial ranges, prominent edges or features in the images, or in case of a LIDAR a reflected ray carrying the spatial information including a certain intensity. These Detections may be best described as "information with no past or future" generated in every cycle of the sensor. The temporal aspect is introduced by the trackers collecting these fragments of information over time and merging them into **Objects** using physical models, carrying a reference to the incorporated Detections.

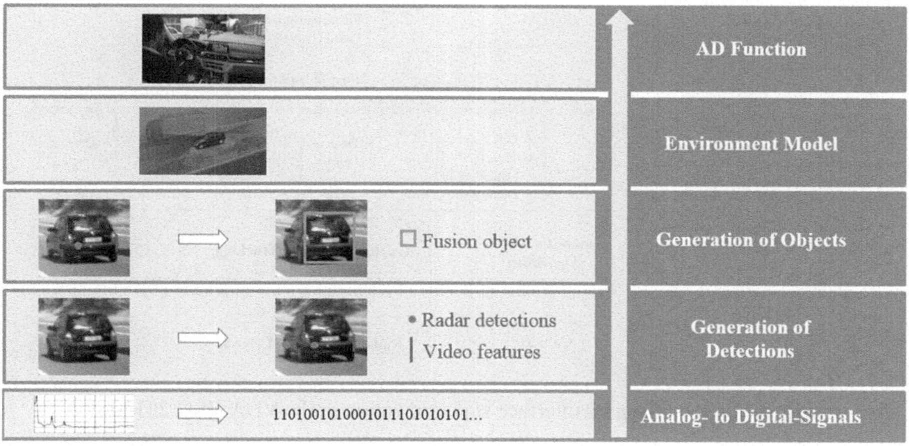

Fig. 2. Generic signal flow of AD systems

The logical sensor interface with Detections and Objects is subject to the ISO standardization initiative in the VDA (2a in Fig. 3) and are described in detail in [1]. It was stated before that a tracking algorithm within the sensor solely using its own information results in Objects with lower quality due to the above described technology restrictions. The following fusion unit can compensate for this and is able to merge the information on two different levels, using Object Level fusion as well as Feature Level fusion [1]. In order to perform Object Level fusion from two different sensors, each sensor tracker needs sufficient evidence in the form of correlated Detections belonging to this particular object in the first place. An advantage of Feature Level fusion is the possible creation of an object in the common fusion unit out of Detections coming from various sensors, which alone would not have resulted in an object detection within the sensor itself [5]. In contrast, fusion on an Object level is less CPU-intensive and mirrors robust state-of-the-art fusion technology [6].

The interpretation of the situation in an **Environment Model,** by giving meaning and importance to the objects and deriving an action resulting in commands to the actuator units of the car, is afterwards handled in the **AD function**. These are mostly developed by the OEMs as they are unique selling proposition of the brand.

It is crucial to note that there are hybrid versions of this architecture. The above describes implicitly that the sensors are provided by suppliers of the OEMs which handle the fusion and function development by themselves. Though specific Euro NCAP relevant functions are possibly integrated on the sensors control unit. An example could be the AEB Interurban System [7], which is running on the camera control unit and performs a simple Object fusion on its own by also considering Objects from a Radar unit prior to the scene interpretation. Since the focus of the VDA standardization initiative is on complex hardware setups for AD functions, an architecture including specific function signals is, in contrast to the fused object list, out of scope.

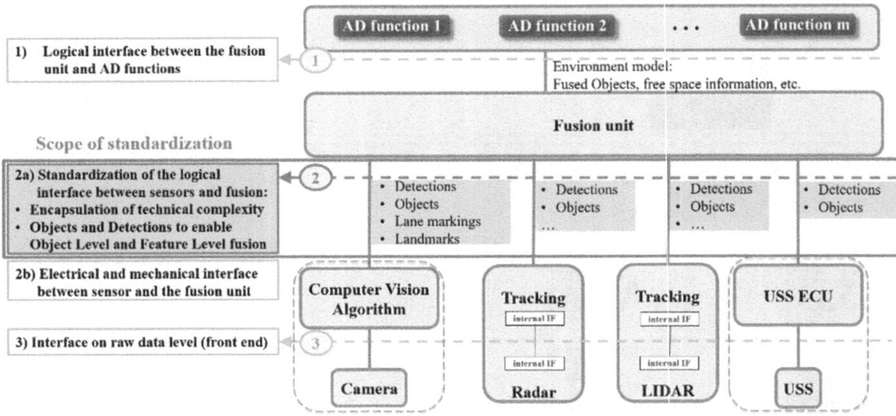

Fig. 3. Scope of sensor interface standardization in the VDA (ISO 23150)

1.3 Status of the Current VDA Initiative to Standardize Sensor Interfaces in ISO

An ISO NWIP (New Work Item Protocol) in regard to standardized sensor interfaces (ISO 23150) was submitted by the VDA Ad-hoc working group "AK Sensorschnittstellen" by industry-leading OEMs and suppliers. Currently the draft document is registered as new project in the ISO committee's work programme (stage 20.00), and therefore in the near future, the first work group meeting will be set up. Interested OEMs and sensor suppliers are encouraged to contribute to the design for standard sensor interfaces with us.

2 Standardized Interfaces in the Context of Virtual Test and Validation

The increasing complexity of AD functions results in the situation that an a-priori definition of test catalogues based on requirements and subsequent field operational tests (FOT) or naturalistic driving studies (NDS) are not covering the necessary test space of millions of kilometres in order to prove the functions safety anymore. Therefore, different test instances like Vehicle in the Loop (VIL), Hardware in the Loop (HIL) and, most importantly at early development stages, Software in the Loop (SIL) are needed in combination with a scenario based approach in order to expand and transfer the test space into the virtual world [8].

Using virtual test and validation at early stages of the development process replaces most parts of the chain of effect with virtual models, because hardware components are available at future dates. Furthermore, virtual test instances can be scaled at lower costs and a single function can be put to test in arbitrary virtual scenarios in parallel.

Picking a lane keeping assistant (LKA) as a visual example simplified in Fig. 4, a developer fuses the lane information from a front camera with the high precision map data using the cars position to form the input for the control system of the LKA. With this virtual test setup, functional tests can be conducted on the side of the control system for example integrated in the build system covering the basic logic of the LKA. In further stages, a mature implementation is also deployed to the control unit of the target hardware and the artificial map data is replaced through the actual interface and map data provided in the car. Moving from a virtual test instance to the target hardware is a continuous process and not mutually exclusive. Considering this, the definition of a common standard for the logical interface between sensors and the fusion unit must also be described by a continuous process aligning the development of virtual replacement models such as sensor models as well as the actual hardware in order to provide a seamless integration.

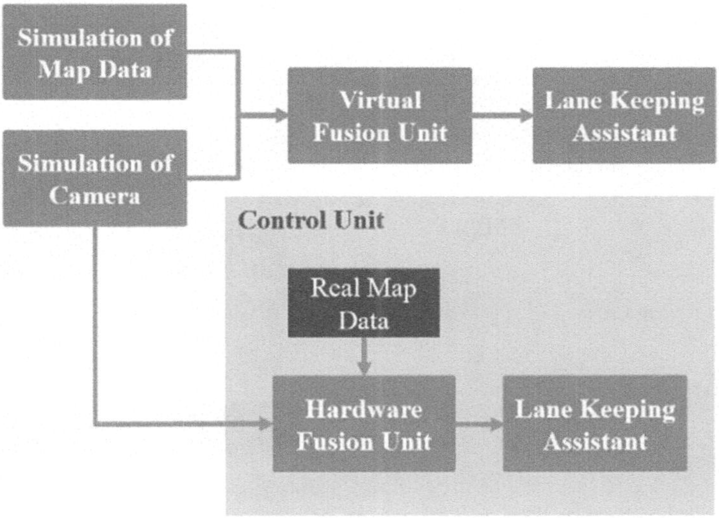

Fig. 4. Lane keeping assistant on SIL and HIL

The Open Simulation Interface (OSI), endorsed by the Pegasus project, is developed on GitHub under Mozilla Public Licence 2.0 as the first reference implementation for the upcoming ISO standard [9]. The main goal is to provide a generic interface for the connection between the function development framework and the simulation environment, while also proposing a standardized way of integrating sensor models using the Functional Mockup Interface (FMI) [10] in combination with the protocol buffers library developed and maintained by Google [11]. The interaction between a serialization protocol like protocol buffers and FMI allows the description of complex data structures and datatypes which can be streamed into a Functional Mockup Unit (FMU) by using a single input byte array terminated by length analogously to a string with "\0" termination which is officially supported with the upcoming FMI version 2.1.

An overview of a closed loop architecture using the Open Simulation Interface for the integration of sensor models can be seen in Fig. 5, describing the integration of sensor models in detail:

Starting with the **Environment Simulation**, a virtual representation of the current simulation state at a single point in time is created using the virtual map data and a scenario description as input. The **osi::GroundTruth** (GT) interface gives an exact view on the simulated objects in a global coordinate system. This interface has no representation in the real world because it describes an idealized state lacking any uncertainty or error aiming to contain all relevant information to model the actual behaviour of an environment sensing sensor.

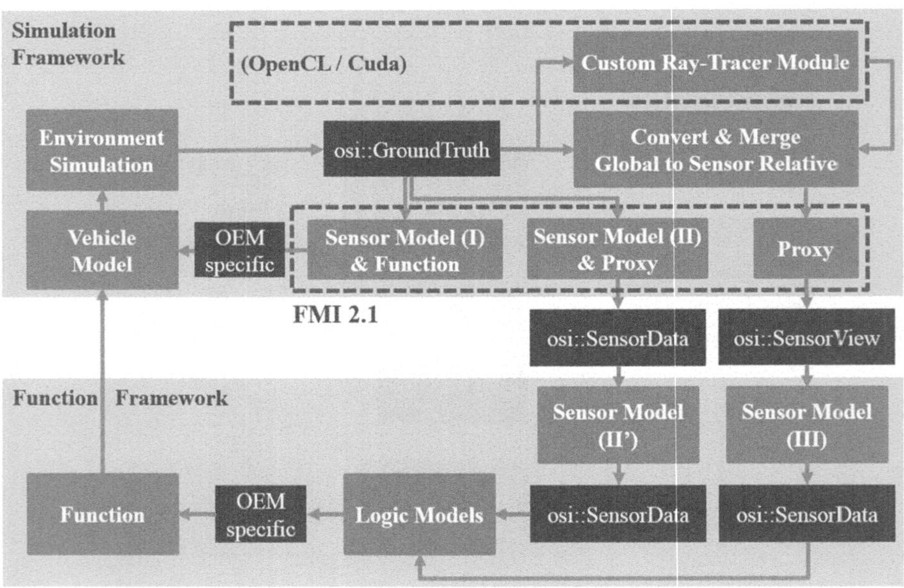

Fig. 5. Closed loop architecture using sensor models for test and validation with OSI

This message is populated using the internally available data and then published to external subscribers through a FMU running within the driving simulation framework. This unit can for example encapsulate a model of a camera sensor converting the GT message to **osi::SensorData** (SD). SD represents the current state of the ISO standardization and is therefore the virtual representation of the actual hardware sensor output including Objects and Detections. This data structure is described in the reference frame of the sensor including the error on the data introduced by its subjective perception of the environment. An example could be the inaccuracy on the position of a car induced by the limited lateral resolution of camera systems which the subsequent fusion unit expects and is obligated to handle accordingly.

This architecture proposal in loan to [12] distinguishes between **Sensor Models** (SM) which are described in detail in [13, 14] and **Logic Models** (LM). The latter ones existence derives from the fact that OEMs typically request additional features from sensors apart from the standardized content of the ISO proposal. In the simplest form, an OEM orders a sensor according to the standard and the LM performs a one-to-one mapping between the data structure of OSI and the current version of the on-board network structure in the car. In a more complex case, the LM implements a complete function running on the sensors control unit.

The standardization contains Feature Level data, such as point clouds from LIDAR sensors, which is filtered by the trackers and merged into Objects described with bounding boxes used to indicate their maximum extend. Comparing this bounding box with a variety of points reflected from the surface of a former complex structure shows a significant loss of information. In order to replicate the information necessary to adequately model the behaviour of Detections on the Feature Level the additional interface **osi::SensorView** is introduced in the Open Simulation Interface.

It supports the usage of **Custom Ray-Tracer Modules** for example using OpenCL or NVidia CUDA allowing the implementation of physical sensor models [15]. The **Convert & Merge** is a block belonging to the simulation framework which transforms the GT to sensor relative data, making use of the already existing features of most simulation tools like defining the mounting position of a sensor or the Field of View (FOV). In addition the data generated in the custom ray-tracer can be associated with the underlying GT. This sensor specific view can afterwards be further processed in order to model the final osi::SensorData structure.

With this architecture sensor models can be classified in three different categories. Figure 6 provides an overview reaching from Ideal Sensor Models, through Phenomenological Sensor Models, to Physical Sensor Models:

Ideal Sensor Model

- Transforms global GT into sensor relative data
- Extended data sheet
- Filters according to technical specifications:
  - Mounting position
  - Field of view
  - Speed limitations
  - Minimal object sizes
  - Resolution
  - Object classes

Phenomenological Sensor Model

- Extension of the Ideal Sensor Model
- Introduces stochastic effects
  - Jitter on positions
  - Object misses
  - Ghosting effects
  - Detection rates
  - Object certainty
- Signal processing effects
- Weather conditions

Physical Sensor Model

- Not solely based on object lists
- Includes physical models
  - Ray-tracing
  - Wave propagation
- Generates „raw“ data
  - Point clouds
  - Reflection lists
  - Images
- Considers specific geometrical properties of objects

Fig. 6. Different categories of sensor models

This order outlines the depth of modelling involved as well as their application during the development process and usage in different test instances:

- The first category of models can be described as an extended data sheet reflecting the general requirements towards a specific sensor which can be used to optimize the overall sensor setup in a vehicle and represents the first stage sensor model in Fig. 4.
- The Phenomenological Sensor Models are an extension to the idealized modelling approach. A FOV is not a binary decision anymore, leaving the sensor blind at the boundary of the technical specification, but describes it as a stochastic transition. Furthermore, jitter on positions, ghosting effects or object misses are introduced following the typical behaviour of a class of sensors or are even derived from measurement trials. A further developed fusion unit requires this kind of behaviour in order to function properly.
- The Physical Sensor Model has the highest modelling depth and finds its application for example in a HIL test for the LKA. The simulation environment renders the image from the perspective of the mounted camera sensor. In a further step the influence of the lens and the imager itself like distortions or exposure are introduced. The modified image is then ingested by the virtual electronical control unit (vECU) responsible for the image processing.

3 Conclusion

This paper introduced the common effort in the standardization of the logical sensor interface for the hardware component in alignment with the development of virtual replacement models for Software in the Loop test instances and others using the Open Simulation Interface (OSI). Its continuous development process is ideally accompanied with a transparent open source project on GitHub where different stakeholder can contribute and work on the requirements towards sensor modelling in a collaborative way. The task of developing and testing the AD functions of the new generation is a joint task of the industry partners together with universities. In the end, products should differentiate themselves by content rather than by the incompatibility of their interfaces.

References

1. Schaller, T., et al.: Proposal for standardization of sensor interfaces to a fusion unit. In: International Congress ELIV, Bonn, October 18th and 19th (2017)
2. Roethig, P., Van Driesten, C.: Open simulation interface. Open simulation interface – pegasus, 3 May 2017. http://www.pegasus-projekt.info/de/72/open-simulation-interface. Accessed 8 March 2018
3. Holder, S., Hörwick, M., Gentner, H.: Funktionsübergreifende Szeneninterpretation zur Vernetzung von Fahrerassistenzsystemen. In: Automatisierungssysteme, Assistenzsysteme und eingebettete Systeme für Transportmittel (AAET 2012), (2012)

4. Spanfelner, B., et al.: Challenges in applying the ISO 26262 for driver assistance systems (5. Tagung Fahrerassistenz, München, Mai 15th and 16th) (2012)
5. Tanzmeister, G., Steyer, S.: Spatiotemporal alignment for low-level asynchronous data fusion with radar sensors in grid-based tracking and mapping. In: IEEE International Conference on Multisensor Fusion and Integration for Intelligent Systems (MFI), S. 231–237. IEEE (2016)
6. Aeberhard, M., et al.: Track-to-track fusion with asynchronous sensors and out-of-sequence tracks using information matrix fusion for advanced driver assistance systems. In: Intelligent Vehicles Symposium (IV). S. 1–6. IEEE (2012)
7. AEB-Notbremsassistent für höhere Geschwindigkeiten | Euro NCAP. EuroNCAP. https://www.euroncap.com/de/fahrzeugsicherheit/die-bedeutung-der-bewertungen/sicherheitsassistenten/aeb-notbremsassistent-fuer-hoehere-geschwindigkeiten/. Accessed 9 March 2018
8. Groh, K., et al.: Towards a scenario-based assessment method for highly automated driving functions (2017)
9. Hanke, T., et al.: Open simulation interface: A generic interface for the environment perception of automated driving functions in virtual scenarios, 3 February 2017. http://www.hot.ei.tum.de/forschung/automotive-veroeffentlichungen/. Accessed 8 March 2018
10. Blochwitz, T., et al.: Functional mockup interface 2.0: the standard for tool independent exchange of simulation models. In: Proceedings of the 9th International MODELICA Conference, S. 173–184. Linköping University Electronic Press, Munich, 3–5 September 2012
11. Varda, K.: Protocol buffers: Google's data interchange format. Google Open Source Blog. Available at least as early as July 2008, 72. Jg (2008)
12. Hanke, T., et al.: Generic architecture for simulation of ADAS sensors. In: 16th International Radar Symposium (IRS), S. 125–130. IEEE (2015)
13. Hirsenkorn, N., et al.: A non-parametric approach for modeling sensor behavior. In: 16th International Radar Symposium (IRS), S. 131–136. IEEE (2015)
14. Rasshofer, R.H., Rank, J., Zhang, G.: Generalized modeling of radar sensors for next-generation virtual driver assistance function prototyping. In: 12th World Congress on Intelligent Transport SystemsITS AmericaITS JapanERTICO (2005)
15. Hirsenkorn, N., et al.: A ray launching approach for modeling an FMCW radar system. In: 18th International Radar Symposium (IRS), S. 1–10. IEEE (2017)

Virtualization for Verifying Functional Safety of Highly Automated Driving Using the Example of a Real ECU Project

Johannes Wagner(✉), Joachim Löchner, and Oliver Kust

ETAS GmbH, Borsigstrasse 24, 70469 Stuttgart, Germany
johannes.wagner@etas.com

Abstract. Validating and verifying highly automated driving (HAD) systems is a huge challenge due to their complexity, short development cycles, and legal safety requirements. The latter are defined by ISO 26262, a standard for functional safety of automotive electric/electronic systems. ECU virtualization is key to mastering this challenge because it allows to transfer testing tasks from the road to simulations on a large scale. If a data-driven approach is chosen for this purpose, comprehensive, loss-free, and time-synchronized recordings of measurement data are required, which must be gathered during extensive test drives. Huge data volumes are the result. These volumes need to be managed, categorized, and processed in a way that they can serve as an input for simulated test drives later. In case of a model-driven approach, the accuracy of the models of driver, vehicle and environment is crucial to obtain meaningful test results. During ECU virtualization, it is vital that the virtual ECU reproduces the behavior of the real ECU as closely as possible. ETAS ISOLAR-EVE can provide substantial benefit due to its way of virtualization. In the end, the value of virtualization depends on sufficient equivalence of the simulated system behavior with the real system behavior. If this can be proven, simulation and virtualization can minimize the need for expensive prototypes, test vehicles, and test drives, while at the same time satisfying legal requirements. In addition, virtualization allows to keep development cycles short and costs limited. The feasibility of this approach is shown using the example of a real ECU project, for which ETAS has provided tools and consulting.

1 Introduction

Advanced driver assistance systems (ADAS) is currently one of the most dynamic areas in automotive development. A number of reasons is responsible for this trend. On the one hand features like active cruise control and congestion assistance are very attractive comfort functions for vehicle owners and drivers, on the other hand legal regulations increasingly require assistance systems. For example, emergency break assistants and lane keeping assistants have been required by legal regulations for commercial vehicles in the EU since 2013. In the European New Car Assessment Program (Euro NCAP), a passenger car can only obtain the full score of five stars nowadays, if

T. Bertram (Hrsg.): *Fahrerassistenzsysteme 2018*, Proceedings, S. 56–66, 2019.
https://doi.org/10.1007/978-3-658-23751-6_6

an automatic emergency brake system is on board. It is reasonable to assume that corresponding regulations and requirements will be further tightened in the future.

As a consequence, driver assistance systems rapidly find their way from premium vehicles into the high-volume segments. This creates economies of scale that are steadily driving down prices, particularly for environmental sensors based on video, radar, lidar, and ultrasound technology.

This development paves the way for economically feasible autonomous driving (AD) and HAD systems. These systems are considered to be an important contribution to further reducing accident rates, because human beings are still the biggest risk factor in road traffic according to statistics. Therefore, system complexity is supposed to grow continuously providing increasingly more support and comfort to drivers and intervening automatically in a growing number of critical driving situations. All these interventions obviously require control of vehicle dynamics, both longitudinal and lateral and thus have direct influence on driving safety (Fig. 1).

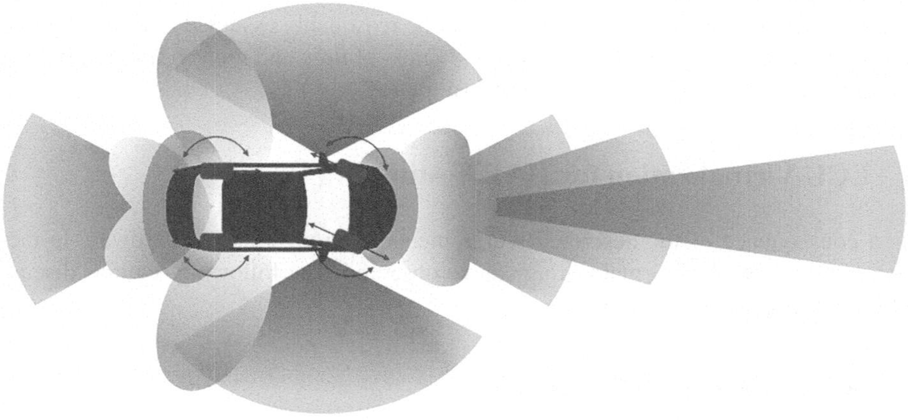

Fig. 1. Sensors and actuators in a system for highly automated driving

2 Legal Safety Requirements

ISO 26262 aims to ensure functional safety of automotive electric/electronic systems by making safety considerations an integral part of each development phase of an automotive product. The standard covers specification, design, implementation, integration, verification, validation, and production release. The standard is an adaptation of the more general IEC 61508 standard "Functional Safety of Electrical/Electronic/Programmable Electronic Safety-related Systems". In the ISO 26262 standard the IEC 61508 standard is adapted to automotive electric/electronic systems. ISO 26262 applies throughout the whole lifecycle of all automotive electronic and electrical safety-related systems.

The ISO 26262 standard defines different levels of functional safety which are based on severity, controllability and probability ("exposure") of potential hazardous events in which a safety critical electric/electronic system can be involved. These levels are called Automotive Safety Integrity Levels (ASIL) and range from ASIL A to ASIL D. Each level is connected with a set of requirements to be met by the development process of the system. ASIL A has the lowest requirements, ASIL D imposes highest requirements and cannot be accomplished without a thoroughly designed processes for both the development itself and the selection and use of development tools.

Since a potential malfunction of an ADAS or HAD system can very likely cause severe injuries or even be life-threatening and will be difficult to control or uncontrollable, ASIL D requirements generally apply to the development of these systems.

System verification and validation are two very important development steps before an automobile can go into production. Even if a system only temporarily takes control of the vehicle, a complete and thorough test is indispensable to ensure that the system works correctly in all possible driving situations and under all possible environmental and vehicle conditions. To take all relevant situations and operating statuses of an ADAS or HAD system into account, a vast number of test kilometers is necessary already today, and the further increasing system complexity will require more and more test kilometers in the future. It can be reasonably assumed that the required test scope cannot be covered within feasible limits of time and money, if only physical test drives are applied for verification and validation.

3 ECU Virtualization for HAD Testing

As a consequence of the described requirements, a strong trend to hardware-in-the-loop testing as well as virtual testing can be observed. The following sections discuss different aspects of this approach.

3.1 Advantages of ECU Virtualization

If testing and calibration tasks can be transferred from the road to simulations on a large scale, they will remain manageable despite increasing complexity and short development cycles. Another advantage of virtualization is the absolute reproducibility of tests. Simulations can be repeated again and again in exactly the same manner. This can even be ensured for dangerous driving maneuvers, for situations at the physical limits of driving dynamics, and for accidents – all of which would be impossible in real test drives. This is why ISO 26262 explicitly recommends simulation for the verification and validation of software safety requirements.

Another benefit of virtualization is that tests can be scaled easily from the individual execution on a single PC to massively parallel testing on a server or even in a cloud. The reason is that the ECUs involved can be both virtually replicated on the PC as well as reproduced indefinitely. Furthermore, the simulation time itself can be reduced, if the virtual ECU allows execution faster than real time. In addition, virtualization allows engineers to work around the clock in teams spread across the globe.

Further advantages include high flexibility regarding the actual test object and testing goal, from the test of individual software components (unit testing), over ECU software integration tests, up to tests on vehicle system level even involving multiple (virtual) ECUs; the short iteration loops between design, implementation, verification, and validation, which help to speed up the process considerably; support for automation, allowing to run nightly tests or continuous (regression) tests in a very effective way; and the direct feedback between development and testing, which directly improves software quality. The latter is particularly facilitated by the fact that virtual ECUs allow software debugging directly on the developers' PCs, and can provide deep insight into internal variables, parameters, and code, in a way that is not easily possible on a production ECU.

All these factors also lead to considerable economic benefits, especially as fewer real test drives and expensive prototypes are then required for testing. Also the early detection of errors and architectural flaws makes a major contribution to cost effectiveness. After all, the later in the development process that an error is discovered, the more complex and expensive it is to correct it.

3.2 Challenges of ECU Virtualization

In the context of legal requirements, three major challenges are connected with ECU virtualization as well.

First of all, the confidence in the overall tool chain needs to be proven. ISO 26262 defines a mandatory procedure for this purpose. Accordingly, each tool that is used in the development process, first needs to be classified based on an application specific analysis of the tool usage and then, depending on the outcome of the classification, in most cases also be qualified for the corresponding usage. Obviously, this demand also applies for the tools that are used for simulation and for virtualization.

Second, evidence needs to be shown that the validated software provides sufficient performance, robustness and effectiveness, not only for the functional aspects of the ECU software, but also for all safety-related mechanisms. Since these aspects cannot all be shown exclusively by means of virtualization and simulation, it can be safely assumed that tests based on virtualization and simulation will have to be combined with tests on the real ECU, e.g. in a hardware-in-the-loop (HiL) setup or in real driving for this purpose.

And third, closely connected with the second challenge, sufficient equivalence needs to be shown between the behavior observed in the simulation and the one of the real system. Again, this is impossible without a certain amount of tests performed on the real ECU. One key method is the comparison between real driving and simulation results. Special techniques can be applied here, which will allow to reduce the tests on the real ECU to the necessary minimum, whereas the large-scale testing of different variants, long-run tests, etc., can be done in simulation. Only this way it will be possible to achieve efficiency improvements in an order-of-magnitude that allows sufficient testing of HAD applications (Fig. 2).

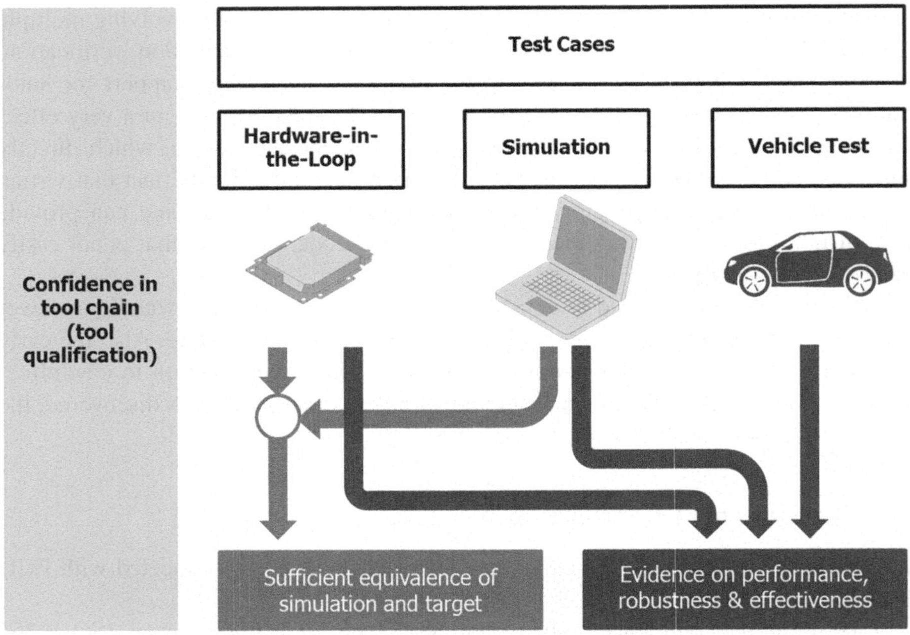

Fig. 2. Challenges of ECU virtualization

3.3 Virtualization Approach for HAD ECUs

To handle both safety requirements as well as functional requirements equally well, hybrid architectures have been proposed for HAD ECUs. They contain at least one microcontroller, which typically features a software architecture based on the AUTOSAR Classic Platform standard. In addition to that, one or multiple microprocessors take care of high performance calculation tasks. AUTOSAR proposes a software architecture for these microprocessors as well, which is described in the AUTOSAR Adaptive Platform standard[1] (Fig. 3).

Virtualizing an ECU, which is based on such a heterogeneous architecture, requires separate consideration of both, microcontrollers and microprocessors, due to their fundamentally different properties. The microprocessor part is often POSIX based and the technical challenge of virtualizing it can be regarded as solved, since POSIX operating systems are regularly supported by off-the-shelf virtualization environments.

The virtualization of the microcontroller constitutes a bigger challenge, since microcontroller architectures demonstrate substantial differences against the PC architecture, on which the virtual ECUs are supposed to be executed. Also the aspect of multi-core processing needs to be observed. ETAS provides a solution, which allows bridging these differences while at the same time maintaining a maximum possible congruence

[1] https://www.vdi-wissensforum.de/news/autosar-for-next-generation-cars/.

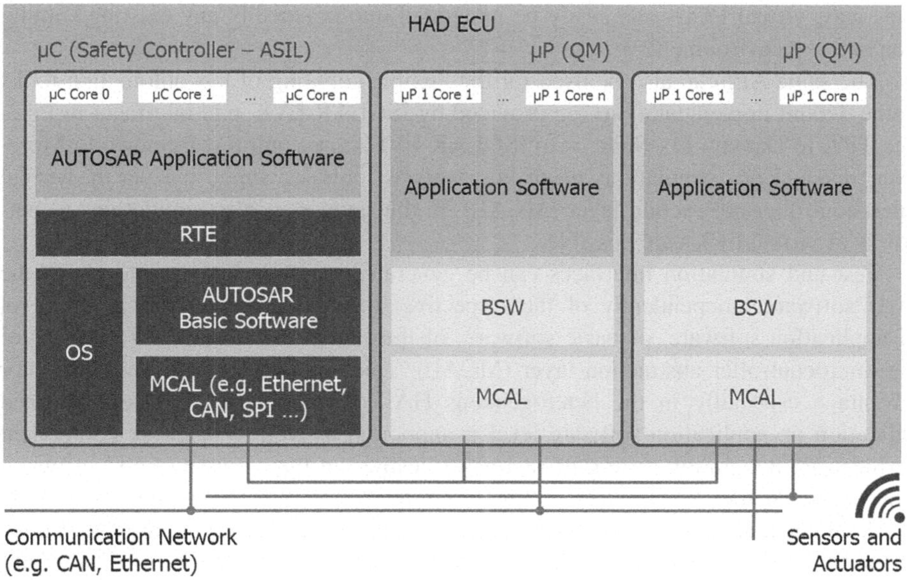

Fig. 3. Possible heterogeneous HAD ECU architecture

between the behavior of the virtualized ECU and the real ECU: the ISOLAR-EVE product.

ISOLAR-EVE distinguishes carefully between two kinds of code, that is to say hardware-independent code and hardware specific code. All code that is independent from the target hardware platform, i.e. application software (ASW), the AUTOSAR runtime environment (RTE), and large parts of the AUTOSAR basic software (BSW), will be used for the virtual ECU directly without changes. The remaining code, i.e. the code that is hardware specific, will be replaced with implementations for the PC that provide comparable functionality. In particular, there are specific implementations of the operating system (OS) and the microcontroller abstraction layer (MCAL), which substitute the corresponding microcontroller-specific implementations. While the OS is based on ETAS' production OS implementation that has been proven in hundreds of millions of production vehicles, the MCAL is specifically designed to use the PC platform in an optimal way. ISOLAR-EVE hence allows to build virtual ECUs, which contain nearly the complete functionality of real production ECUs and thus reproduce their behavior as realistically as possible. This is due to the fact that the virtual ECUs contain not only the ASW, but also nearly the complete BSW and large parts of the production OS – a major prerequisite for the equivalence between the behaviors observed in the simulation and the one of the real system, as required by ISO 26262.

Even better, the resulting virtual ECUs offer comprehensive open, flexible interfaces tailored for the needs of PC-based development and testing. This includes not only a standard Functional Mockup Interface (FMI), an interface to ETAS' co-simulation environment COSYM, and a Simulink® S-function interface, but also the possibility to freely implement nearly any tool coupling over an open .NET API.

This way, virtual ECUs can easily be integrated into practically any existing simulation and test environment.

The same API can also be used for the flexible coupling of test automation tools. Some default implementations are provided by ISOLAR-EVE, e.g. interfaces to PikeTec TPT, to Danlaw Mx-Suite or to ISOLAR-EVE's own unit test framework. Moreover, residual bus simulations using bus analysis tools are supported via the virtual microcontroller abstraction layer (MCAL). In the same way, also simulations of networks of virtual ECUs are possible.

Test and simulation interfaces can be generated at all architecture levels of the ECU software independently of the respective software manufacturer: at the level of application software, of basic software, of the runtime environment (RTE), or of the microcontroller abstraction layer (MCAL). This functional flexibility is a major advantage especially in the heterogeneous HAD development, because functional validation on application software level is supported as well as software component verification, integration testing or regression testing on the complete ECU software. Hence, during the development of highly complex software algorithms for a HAD ECU, an adequate concept for automated testing can ensure at any moment that none of the functionality that has already been implemented will be deteriorated by a later development step (Fig. 4).

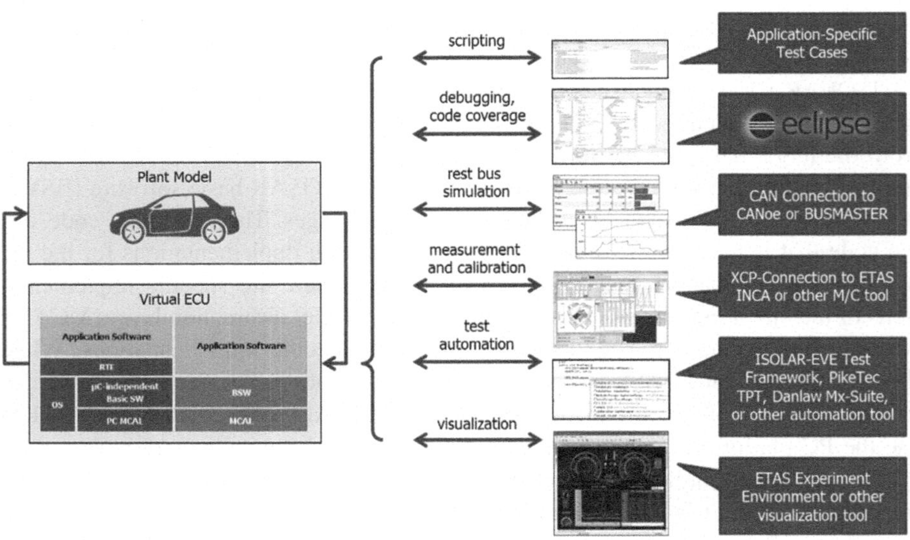

Fig. 4. Open interfaces for flexible usage of virtual ECUs

As ISOLAR-EVE is based on the Eclipse platform, developers can directly reuse existing Eclipse functionality, such as debugging, code coverage analysis, coupling to version management systems and specialized editors.

In many cases, execution speed faster than real time is an advantage of virtualization. Nonetheless, it is sometimes also required to execute tests in hard real time, e.g. when the virtual ECU is combined with specialized hardware devices, real sensor inputs or with a hardware-in-the-loop system. In these cases, virtual ECUs created with ISOLAR-EVE can be run on Linux-based ETAS real-time PCs (RTPCs), which ensure real-time operation and communication.

In the future it will no longer be enough to validate ECUs and their software individually. Instead – and specifically with respect to HAD – proof of functionality at the system level will be needed. This validation at the system level requires integrated test environments into which virtual ECUs can be incorporated in simulations and linked to HiL setups from various vehicle domains. In conjunction with ETAS's multi-RTPC technology, it is possible to operate virtual and real ECUs from various vehicle domains in networked connections with real-time synchronization of data traffic. Such system setups are nearly unlimited in their scalability and cannot only be used for modeling a network of multiple HAD ECUs in one single vehicle. They are even capable of modeling highly complex networks of multiple vehicles with HAD functionality.

4 Simulation Environment

For the representation of the environment, the vehicle, and the driver, there are two major approaches: a model driven and a data driven approach. Each of these approaches is connected with its own challenges, as well.

In a model-driven approach, virtual ECUs are usually executed in closed-loop simulations. Their inputs and outputs are connected with models of the driver, the vehicle including all its sensors, and the environment. For this purpose, simulation models with sufficient fidelity are indispensable, which require profound knowledge of the system details.

In a data driven approach, virtual ECUs are typically simulated in an open-loop setup. Their inputs are stimulated with data representing the sensor measurements, usually from recordings taken during real test drives, and their outputs are compared with corresponding reference values. In this case, major challenges are to obtain, store, and organize the amount of input data required for sufficient test coverage. This data should be not only be recorded across a wide range of driving situations. Moreover, it should be comprehensive, complete, and time-synchronized. An approach to this will be discussed in the following.

4.1 Using Measurement Data as Stimuli

As already indicated before, the success of a data-driven open-loop testing approach is highly contingent on the quality of the data used as stimuli for HiL and SiL tests. Ideally, this data is recorded during real test drives.

On account of the huge amounts of data involved, however, recording requires technology with suitable bandwidths, computing power, and memory capacity. The need is not only to record the data from multiple ECUs in the vehicle but also to record the raw data from the sensors capturing the vehicle surroundings. If multiply redundant 360-degree environment monitoring becomes established as autonomous driving evolves, data quantities will increase massively.

In addition to recording the data, it has to be managed appropriately. To ensure that developers can actually benefit from the data for the test of HAD systems, it must be systematically stored and made retrievable using suitable indexing and search engine technology.

HAD controllers need to react on driving situations, especially when they are critical, within fractions of seconds. Hence, it is indispensable that all data recorded in test drives is temporally precise and completely time-synchronized. This property must be ensured in spite of high data rates, heterogeneous architectures with differing numbers of ECUs arranged in various hierarchies, and a broad range of communication methods (CAN, LIN, and FlexRay bus systems, Ethernet) and data formats. For this purpose, time stamps can be captured to ensure that the data can be synchronously played back later.

Today's vehicles and development tools already provide some prerequisites for these tasks: high-performance processors, data buses, and communication protocols in the vehicle, and high-performance, scalable tools for recording and logging measurement data during test drives. However, their performance capability will have to exceed today's levels if the test vehicle data is to be recorded completely without any gaps. And data management solutions will also have to be optimized to fulfil the tasks mentioned above.

4.2 State of the Art

For the comprehensive and time-synchronized recording of data flows in today's driver assistance systems, first the internal data of all microcontrollers involved must be captured and logged together with time stamps. Second, the raw data from video camera and radar systems must also be recorded with time synchronization. For the controller data, the bandwidths are in the two digits MB/s range. The Ethernet-based ECU interfaces and interface modules currently available meet this requirement, which also means that precise time synchronization based on the Precision Time Protocol in the open IEEE-1588 Ethernet standard can be guaranteed as supported, for example, by the latest generation of ETAS hardware devices. The time synchronization is accurate down to the nanosecond for hardware and to a few microseconds for software.

Recording the raw data from environmental sensors is much more demanding. Taking current front camera technology as an example, the data streams supplied by the imagers come in bandwidths of around 100 MB/s. The hardware needed for this purpose generally exceeds the capacity of standard products, as the requisite infrastructure is far more complex than the one required for recording controller data. For the latter, the ETAS microcontroller access technology, ETAS ETK, can be used, which has already become established for powertrain control. Thanks to its modular design and standardized interfaces, ETAS ETK technology is also scalable for the much higher data traffic of a heterogeneous ADAS system environment.

4.3 New Requirements for Data Acquisition

For higher degrees of automation, new laser-based lidar sensors supplement the camera, radar, and ultrasound systems already in use, making the monitoring of vehicle surroundings both near and far ever more closely meshed. Highly complex domain ECUs will be needed, which execute arithmetic-intensive algorithms to handle the resulting amounts of data, e.g. for sensor fusion. As already described, these ECUs will feature heterogeneous hardware architectures consisting of microprocessors and the rather traditional microcontrollers, making new demands to data acquisition as well. ETAS has developed new, highly innovative solutions to satisfy these demands and to allow for data acquisition also on such new architectures.

In addition to the required sensor redundancy and the introduction of domain ECUs, additional factors indicate that data traffic will rise further in the future: camera resolution is increasing continuously and there is a trend from grayscale to color shade capturing. On the one hand, this requires the development of hardware capable of handling data in the requisite bandwidths of several GB/s on a time-synchronous basis. On the other hand, data compression will be inevitable, whereby it must always be ensured that the computing power required for this data compression neither affects ECU functions nor corrupts measurement data.

4.4 Systematizing Captured Measurement Data

When all signals from vehicle electronics are completely recorded, data quantities will easily enter the petabyte range. To create an overview in such extreme data amounts, new tools are also needed: intelligent operating concepts, extremely flexible database technologies, powerful search algorithms along with navigation and selection mechanisms, and visualizations that help keep track of things.

A broad field for innovations opens up here too – innovations that will allow the complete, gap-free measurement data recordings from test drives to be reused again and again for new purposes and unlock the full potential of the growing data stock. Various projects are working on the application possibilities of existing big data technologies in the automotive area.

ETAS already offers a range of big data solutions specifically for the capture and central storage of ADAS data as well for the indexing and flexible linking of the data for search functions and for analysis and reporting.

5 Example

Together with a large German ECU manufacturer and a premium vehicle manufacturer, ETAS is working on the virtualization of an ECU for partially automated driving. This ECU features a heterogeneous architecture consisting of one microcontroller and one microprocessor, each with multiple cores. They are virtualized with ISOLAR-EVE in combination with regular off-the-shelf virtualization software. Different simulation setups are used. One of them is a closed-loop setup for function validation on pure application software level using simulation models of the driver, the vehicle, and the environment. Another one is the software verification by means of open-loop simulation based on recorded data. Large parts of the simulation and test environment have been available at the vehicle manufacturer already. ISOLAR-EVE's open interfaces were used to establish the simulation and test interfaces. In addition to that, communication among the virtual microprocessor and the virtual microcontroller and to rest bus simulation and test tools is established on MCAL level over several communication interfaces.

One focus of the implementation has been to enable ISO 26262 conform software verification based on virtual ECUs and simulation. ETAS has advised both parties regarding the safety concept and details of the ECU configuration and implementation, e.g. on timing aspects.

Thanks to virtualization, the validation of ECU software functions, the interaction of micro-controller and micro-processor, and their interaction in the network could begin long before availability of corresponding hardware prototypes.

6 Conclusion and Outlook

ECU virtualization helps engineers to manage the increasing complexity and safety requirements that arise from the development of driver assistance systems and highly automated driving systems. It makes debugging, validation, and verification much more effective and is an enabler for achieving the necessary test coverage required by safety standards as ISO 26262. At the same time, virtualization helps to keep development cycles short, not only because test activities can begin far before the availability of ECU hardware, but also because the operation of virtual ECUs can be scaled easily form a single developer's PC up to massive parallel testing on a server or in a cloud. This leads to significant economic benefits. Moreover, fewer real test drives and expensive prototypes are then required for testing, facilitating additional cost reductions.

However, virtualization does not come for free: recording and management of comprehensive test data of high quality, the transformation of the ECU software into a virtual ECU, and the required fidelity of simulation models, are connected with several challenges that engineers need to cope with. After all, virtualization only provides the intended value, if sufficient equivalence between the simulation and the real system can be achieved and proven. Virtualizing nearly the complete ECU software, as ETAS ISOLAR-EVE does it, is an ideal foundation for this purpose.

This paper shows at the example of an ECU for partially automated driving that ETAS tools and consulting expertise can help mastering these challenges. It introduces concepts that are also sustainable for the development of HAD ECUs. ECU virtualization and mastering large data quantities are keys to managing the extensive testing required to verify and validate HAD systems despite ever faster development cycles.

In the future, with regard to autonomous driving, comparisons between the behavior of a HAD system and human test drivers can even pave the way for the use of deep-learning algorithms, virtual sensors, and further methods from the domain of artificial intelligence.

Derivation and Application of an Observer Structure to Detect Inconsistencies Within a Static Environmental Model

Moritz Lütkemöller[1(✉)], Malte Oeljeklaus[1], Torsten Bertram[1], Klaus Rink[2], Ulrich Stählin[2], and Ralph Grewe[3]

[1] Institute of Control Theory and Systems Engineering, TU Dortmund, Dortmund, Germany
moritz.luetkemoeller@tu-dortmund.de
[2] Continental, Chassis & Safety, Passive Safety & Sensorics, Frankfurt, Germany
[3] Advanced Engineering, Continental, Chassis & Safety, Frankfurt, Germany

1 Introduction

1.1 Motivation

Smart vehicles like autonomously driving cars have the advantage to make a ride more safe and comfortable because intelligent vehicles respond better and faster than human beings to critical situations [1]. Having accurate knowledge of its environment is essential for a car that drives autonomously. Based on this, a maneuver adapted to the situation can be planned and carried out. The perception takes place via a variety of sensors such as cameras, radars and LIDAR's [2, 3]. However, the disadvantage of an intelligent vehicle is the complex infrastructure consisting of sensors, ECUs, communication equipment, etc. that is needed to perceive the environment correctly. As the complexity increases, the susceptibility to errors of such a system increases the same way. Errors refer to the failure of individual units, the incorrect processing of information, but also the manipulative interference of unauthorized persons.

Within this paper, we discuss a novel approach to assess the consistency status of a static environmental modeling (EM) process. Therefore, the idea of a consistency observer (OS) is presented. In this case, the observer is not a utilized as usual like Luenberger [4] of control theory. Instead, the structure uses methods from the field of machine learning algorithm. The requirements for the OS are avoidance of mathematical and physical redundancy, avoidance of signal reconstruction, a simple transfer to other signal processing systems and a hard decision whether the corresponding environment model is consistent or not. Furthermore, the OS should not depend on the knowledge of experts or, in other words, the system under observation should be considered as a black box.

T. Bertram (Hrsg.): *Fahrerassistenzsysteme 2018,* Proceedings, S. 67–79, 2019.
https://doi.org/10.1007/978-3-658-23751-6_7

1.2 State of the Art

Error and Fault Detection. According to [5] error and fault detection and diagnosis techniques can be divided into three classes

1. Qualitative models and search strategies
2. Quantitative model-based methods
3. Process history-based methods

The first class assumes that there is detailed knowledge about the physical and mathematical relationships of the process that is to be monitored. The search for faults and errors is performed using previously established rules and, in contrast to the quantitative model-based methods, not based on physical signal simulations. As an example [6] distinguish between a topographic search and your symptomatic search. Topographic searches perform malfunction analysis using a template of normal operation. Symptomatic searches instead look for symptoms to guide the search to the fault location. Symptoms need to be known to find them. Examples of this diagnosis class are fault trees [7].

The second class named quantitative model-based methods deals with a model parallel to the process to reduce errors in the process from the difference between the two output signals. The model can be different in nature. On the one hand, it is possible to model the internal process operations in different levels of detail. This is also called analytical redundancy. This principle is primarily used for physically and mathematically well-described systems like mechatronic systems. Furthermore, state reconstruction by, e.g., the Luenberger observer can be done to check the range of certain states. This principle is used, for example, in [8]. In this case, a non-linear aircraft model was used in parallel to monitor an aircraft to conclude from the conditions on defects in the wings. [9] use an Unknown Input Observer (UIO) to detect faulty sensors, which can be used to diagnose both single sensor noise and simultaneous interference in multiple sensors.

In comparison to model-based methods, process-historical methods are based, on the qualitative or quantitative a priori knowledge of the process, but in this case only on a large amount of data that comes from the process. The features of these data can be used to generate knowledge of the process either by experts like qualitative model-based approaches or by machine learning techniques. Among the qualitative methods, [10] have used the concept of frequency spectrum analysis to predict possible malfunctions successfully of a rotor by comparing a malfunction signal and a normal signal with a measurement.

Furthermore, [5] proposes machine learning like neural networks as process modeling. Similar to parallel physical models, errors can be detected by comparing the output of both models. Also, an analysis of process statistics, represented by handcrafted features, can be evaluated and classified by machine learning techniques.

Experimental System. Subject and basis of this work is the EM developed by Continental Teves. The static environment model is analog to the previous chapter represented by an occupancy Grid as presented by [11]. Figure 1 shows the schematic structure of static environment modeling. The grid is formed by the information fusion of the radar sensor on the vehicles' front and the stereo camera, which is located on the rearview mirror in the windscreen.

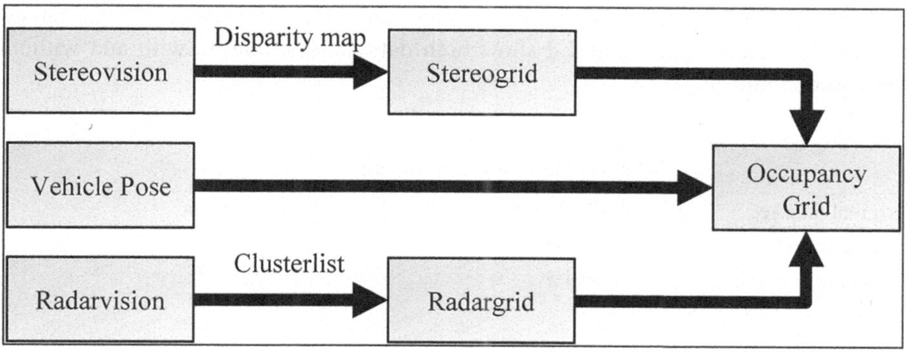

Fig. 1. Schematic overview environmental model

The stereo camera provides a 512 × 220-pixel disparity map $img_{disp}(x, y)$. It is calculated by the disparity of two simultaneously taken images and represents the depth of each pixel along the x-axis in vehicle coordinates.

The next step is the calculation of a grid-representation $Stereogrid(r, \varphi)$. Different to the disparity grid it is a top view (x-y-plane). The resolution of the grid is $-25° < \varphi_{camera} < 25°$ with $\Delta\varphi = 0.1°$ and $2.2\ \text{m} < r_{camera} < 35\ \text{m}$ with $\Delta r_{camera} = 0.1\ \text{m}$ with respect to sensor coordinate frame. Each cell has a defined state, either free, occupied or unknown and contains a value in between 0 and 255 as a probability for the respective state. The grid is not accumulated over time.

The radar sensor provides cluster points C_R. Each cluster is an accumulation of radar reflections and contains information about its distance to the sensor origin in Cartesian coordinates. Particularly as processing the disparity map, a grid $Radargrid(r, \varphi)$ is built. Again, each cell of the grid has the state free, occupied or unknown and a probability for the respective state in-between 0 and 255. The resolution of the grid is $-30° < \Phi_{Radar} < 30°$ with $\Delta\varphi_{Radar} = 0.1°$ and $0\ \text{m} < r_{Radar} < 30\ \text{m}$ with $\Delta r_{Radar} = 0.1\ \text{m}$. Due to the sensors mounting points both grids have nearly the same field of view.

In the last step, the grids are merged into the Occupancy grid $DstMap(x, y)$ following the Dempster Shafer theory of evidences [12]. The occupancy grid is represented in world coordinates and consists of 512 × 512 cells, each 0.16 × 0.16 m. The ego-vehicle is in the middle of the gird. Each cell gets an evidence for being free, being occupied and being unknown. Each evidence is between 0 and 16.

The mathematical procedures for fusion are not considered at this point since they are not relevant due to the OS's requirements regarding the system being monitored as a black box.

Dataset and Inconsistencies. The dataset consists of 1.5 h or 75000 data points of highway driving scenarios. The records were made at sunny and rainy weather along with normal traffic conditions. Since the data record has only a very small proportion of inconsistencies, in addition to these, hypothetical inconsistencies are also manually fed into the data record. This leads to an increased data set of ~90000 data points. The sections of consistent and inconsistent parts within all measurements are labeled by hand. To avoid blurry labeling, the data points at the transitions between the classes have been deleted. Tables 1 and 2 show both datasets distribution with and without handcrafted faults.

Table 1. Distribution of the original dataset

Original data set	Consistent	Inconsistent
# Feature vectors	70588	576
Percentage	99.1%	0.9%

Table 2. Distribution of the modified dataset

Modified data set	Consistent	Inconsistent
# Feature vectors	66348	19,266
Percentage	78.3%	21.7%

An inconsistent EM is defined as a faulty representation of the environment by a different perception of both sensors. Likewise, an intrusion of unauthorized persons or software is seen as an inconsistency. A situation within heavy rain deals as an example for an encountered inconsistency within the recorded dataset. Here, the stereo camera did not recognize the bridge pillar on the right side of the lane while the radar did. Within the occupancy grid, the area is marked as free. In case of a planned trajectory within this area, it would lead to an accident. As an example for a hand-crafted inconsistency, the disparity image interfered with Gaussian white noise. This also leads to a wrong perception and fusion of the occupancy grid.

2 Observer Structure Derivation

When choosing an OS, it is essential to avoid an exact replication of the existing environment model or single signals. Redundancy would lead to low profitability, as there would be need of multiple ECUs. In addition to increased energy requirements, this would result in increased production costs and vehicle weight. At the same time, holistic monitoring is getting even more difficult. Therefore, a classical observer of control technology, as described in Sect. 1.2, falls out of the selection as a physical replicating structure. Also, it is physically not possible to reconstruct the EMs states “consistent” and “inconsistent” as an observable state in the system. Therefore, a classical observer does not meet the requirements.

In the same way, hardware redundancies both diversified and homogeneous, do not meet the requirements. This should also be avoided for economic reasons as told before.

Instead, the OS should be a higher authority, that regards the system to be observed as one big black box model or several composed black box models – a process that can be viewed without any knowledge of its internal workings regarding its outputs and inputs. Only defined intermediate process variables and their characteristics respective features are available for observation. This approach makes it possible to reduce the system complexity since not every in the system existing and perhaps unnecessary information is used. Furthermore, this view of the process along with machine learning techniques leads to high generalizability and transferability of the concept to other parts of the environment monitoring, for example, the dynamic obstacle tracking. Therefore, it is important that the OS can be extended as efficient as possible to new training units and signals. In general, it is relevant to be able to ensure the applicability to an ECU, which is usually associated with limited resources. A high degree of parallelizing of the tasks, such as a mathematical less complex computations, is important to be able to keep the OS's task as centralized as possible even on an ECU with low hardware performance. It is also important that the OS's output is easily interpretable. Ideally, the output only distinguishes between a consistent or inconsistent EM to ensure whether it is usable for e.g., ADAS or not. Also, the OS must be abstract to be able to transfer the concept as easily as possible to other functions or control devices that need to be monitored. It is advisable to keep the level of detail of the structure as low as possible. Thereby, the advantage causes that resources for operating the OS are kept low.

This leads to a statistical classifier along with feature engineering as a proposed OS. It meets the requirements by avoiding signal reconstruction as well as process modeling. Instead, features from input and output signals of the EM processing steps are used to train a supervised classifier. Also, there is a good generalization and transferability of the concept, because an existing training data set can be extended successively extended with test cases or can be replaced by other data sets. In case of an extension of the OS to further processes, only feature engineering needs to be done. Figure 2 shows a schematic view.

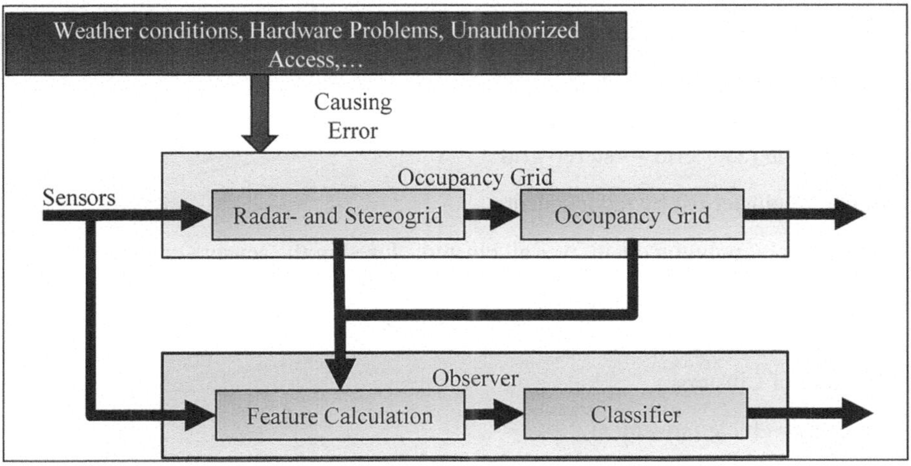

Fig. 2. Schematic overview observer structure

2.1 Feature Engineering

The consistent operation is characterized mainly by given relationships between vehicle speed, stereo grid, radar grid and occupancy grid. Due to the same field of view, radar grid and stereo grid should be constituted similarly. Furthermore, a correlation between current and former measurements, especially within the grids should be given. The mean delta between two measurements is 80 ms. Further, the number of cells classified as free and occupied should not vary much. Within the occupancy grid, cells can be marked as free and occupied at the same time, even with the same evidence for each. The measurements show that this behavior seems to be normal for a few cells. Also, every rate of change within the grids should be proportional to the vehicles speed and yaw rate. These relations between all values are represented within the following features. These features will be used to train and test a supervised classifier:

Features calculated for occupancy grid

- Correlation of OG(t) and OG(t – 1) with respect to v and φ
- Number of cells in conflict states

  › $E_{\text{occ}}(OG(x_i, y_j)) > 0 \wedge E_{\text{free}}(OG(x_i, y_j)) > 0 \; \forall i, j \in OG$

  › $E_{\text{occ}}(OG(x_i, y_j)) = E_{\text{free}}(OG(x_i, y_j))$ with $E(OG) > 0$

- Number of free/occupied/unknown cells
- Mean and variance Evidence over all cells
- Area marked as free in m^2

Features calculated for radar grid and stereo grid

- Correlation of *grid(t)* and *grid(t – 1)* with respect to v and φ
- Number of free/occupied/unknown cells
- Mean and variance Evidence
- Free area in m^2

Subsystem spreading features

- Correlation radar resp. stereo grid ↔ Occupancy Grid
- Correlation radar grid ↔ stereo grid

Features for cluster list by radar sensor

- Free Area by polygon chain over all clusters closed in the sensors origin

$$A = \frac{1}{2}\left|\sum_{i=1}^{n} x_i * y_{i+1} - y_i * x_{i+1}\right| \text{ in } m^2$$

- Number of Clusters

Features for vehicle movement

- Velocity in $\frac{m}{s}$
- › Alteration rate of vehicle in $\frac{rad}{s}$

Features for disparity map

- Correlation *D(t)* and *D(t – 1)*

Furthermore, the difference between each timestamp values is used as features. This leads to 38 features per data point.

2.2 Classifier

[13] has examined the use of bootstrap aggregation (bagging) and boosting techniques for ensemble learner random forest regarding noisy and highly unbalanced data. [14] introduced random forest as an ensemble learner method for regression and classifying. A multitude of decision trees is trained along the dataset and the output class that is the mode of the classes (classification) or mean prediction (regression) of all individual trees. Boosting and bagging are used for reducing bias and variance of a classifier by combining a number of weak learners, decision trees to one strong [15]. Each tree is trained with a new subset of the dataset. Boosting [16] weights the probability of each sample. Therefore, some of them will take part in the new sets more often. The weights are redistributed after each step of training in a way, that misclassified data increases its weights to emphasize the most difficult cases. While Bagging [17] takes place, any element has the same probability to appear in new subsets. The points of data that do not appear in the subset are used for an evaluation of the tree; so called Out-of-bag error (OOB). Variations of Random Forest are Boosting and Bagging as explained in, e.g. [18, 19].

[13] has shown that the methods of bootstrap aggregating have an advantage over boosting. For this purpose, the authors applied various techniques of the boosting method in the form of the synthetic minority oversampling technique (SmoteBoost) and Random Undersampling Boosting (RUS). Both techniques were used once with resampling and reweighting.

Bagging techniques include Exactly Balanced Bagging (EBBag) and Roughly Balanced Bagging (RBBag), each once sampling with and without replacement. This refers to the bootstrap coming out of the amount of training data that is pulled to train a decision tree. The test data came from [20]. The data was converted into a 2-class problem with an imbalance of about 97% to 3% was created. The authors recommend a bagging algorithm that is trained by sampling without replacement. There is no significant difference between the use of EBBag and RBBag.

[21] shows, however, that as the size of the data set increases, the differences between the two methods become smaller. Now, the OS's dataset is already about nine times as big as [13]. Further, a major enlargement is suspected. As a result, the classifying part of the OS will be implemented in the form of random forest with bootstrap aggregating, whereby the bootstrap will be created by using sampling with replacement. That causes the advantage that there is at least one less parameter, which could simplify possible adaptability especially in continuing that work. Furthermore, easy evaluation of training is possible due to the Out-of-bag error calculation that results from training.

3 Observer Structure Application

The dataset is subdivided into 85% training and 15% test data. The split takes place randomly for each class except the not hand-crafted errors to make sure those appear in both either training and test. Furthermore, test data was balanced due to an easier evaluation. The original non-manipulated measurements are not used for training or testing. At first, a classifier is trained with the full amount of features presented before. Afterwards, an evaluation takes place followed by training with a reduced feature set concerning the importance of the feature. The class "inconsistency" is similar to "positive" and "1", while "consistent" is similar to "negative" and "0".

At first, the amount of decision trees is determined. Therefore, we trained 30 random forests with each 200 trees. Figure 3 shows that the OOB-error and its variance converges at about 50 decision trees.

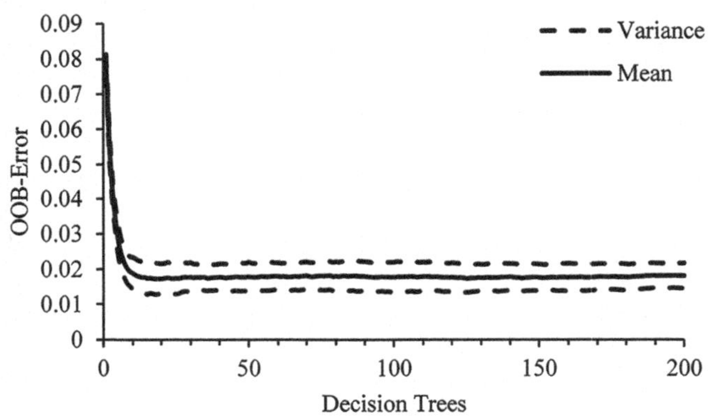

Fig. 3. Out of bag error for different amount of decision trees

The OOB analysis of the 50-tree-trained classifier is shown in Figs. 4 and 5. Figure 4 left-handed describes the number of misclassifications of measurements within training, called outliers. Ideally, no measurement has outliers. The worst-case scenario is a record that consists solely of outliers, so classification is not possible. In this case, there is a purely random process. It can be seen that the amount of outliers per observations decreases logarithmically. At the same time, it can be seen in Fig. 4 right that the middle margin of the classifier converges to 0.95 already after the twentieth trained tree.

The optimal value would be one. Furthermore, Fig. 5 left shows, that the receiver operating characteristic is almost at its optimum. The optimum is a step. Finally, Fig. 5 right shows that the optimum of classification accuracy is achieved if the majority threshold for the class inconsistency is reduced to 30%.

The classification results of the test data can be found in Table 3. 590 of situations classified as inconsistent are in fact consistent, corresponding to a false positive rate of 4%. Conversely, 1107 feature sets are incorrectly marked as consistent, even though this is not the case in reality. This results in a true negative rate of 7.4%. Within the entire number of 5017 vectors that are classified as not usable, 91.1%

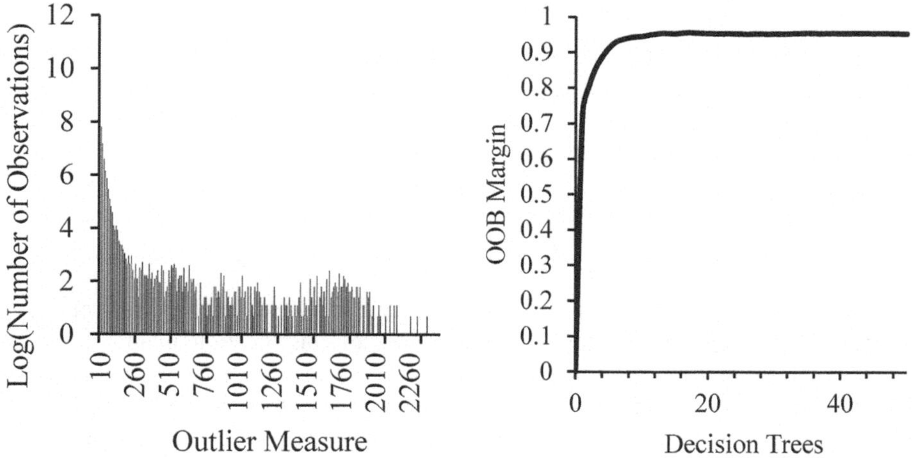

Fig. 4. Full amount of features left: outlier measurements, right: out of bag margin

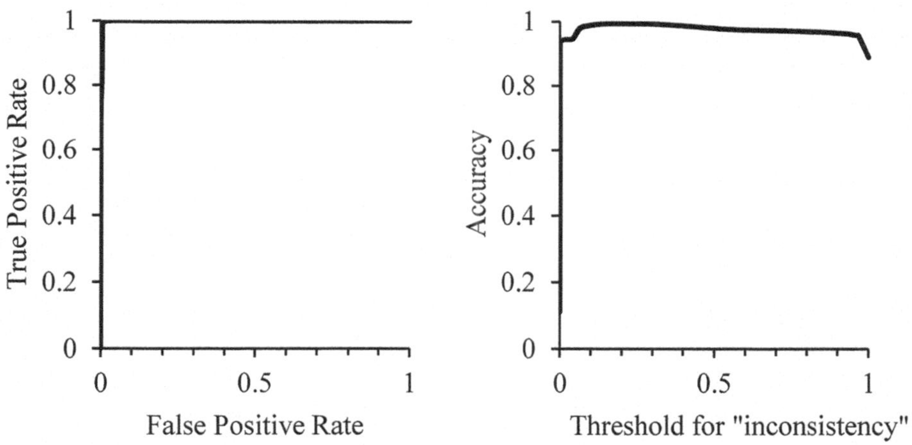

Fig. 5. Full amount of features left: receiver operating characteristics, right: threshold for votes for class "inconsistency"

Table 3. Confusion matrix observer structure full amount of features

	EM inconsistent	EM consistent	
OS inconsistent	40.6%	4.0%	91.1%
OS consistent	7.4%	48.0%	86.6%
	84.5%	92.4%	88.6%

(positive predictive value) are correctly classified and 9.3% wrong. Within the number of 9869 feature vectors that are classified as usable, 91.1% are correctly assigned (negative predictive value), and 13.4% are incorrect. The sensitivity is 84.5% and the specificity 7.6%. The values were calculated according to Table 4.

Table 4. Confusion matrix explanation

	EM inconsistent	EM consistent	
OS inconsistent	Right positive	False positive	Positive prediction value
OS consistent	False negative	True negative	Negative prediction value
	True positive rate	True negative rate	Positive likelihood ratio

All in all, 88.6% of the tested situations are assigned to their correct class, whereas 11.4% are misclassified. Problems and misclassification mainly occur in measurements that are at the transition of classes. Thus, a large part of the false positive classifications comes from a test drive in the rain, in which the EM is available to a limited extent, but is still functional. The correct-negative cases, in turn, arise mainly in the classification of artificial inserted negative classes with the detection of a noisy camera image for the largest part of the cases. A further improvement of the classification result, as well as the calculation performance, is achieved by the reduction of the used features. To comply this, Fig. 6 deals with the OOB error with regard to the number of the most important features. These are determined by the OOB analysis. It can be seen that a reduced amount of 29 features is sufficient for converge of the OOB-error.

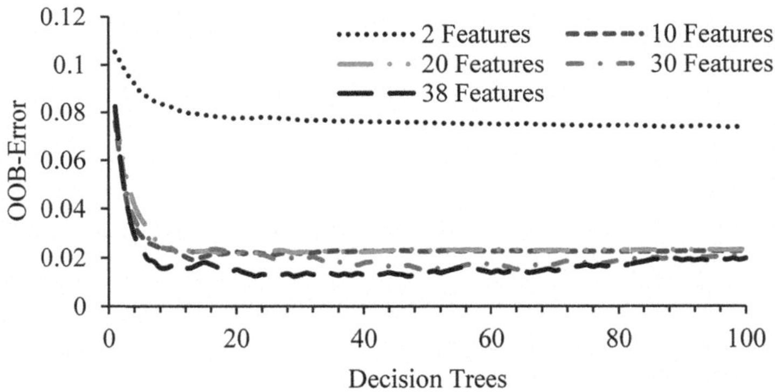

Fig. 6. Out of bag error with respect to different amount of features

The OOB analysis within Fig. 7 of the classifier trained with 50 trees and 29 features shows that there are no significant differences from the classifier described before. Outlier measurements, as well as OOB margins, do not have any appreciable differences, same for ROC. Again, the threshold for the majority is set to 30% for class "inconsistency".

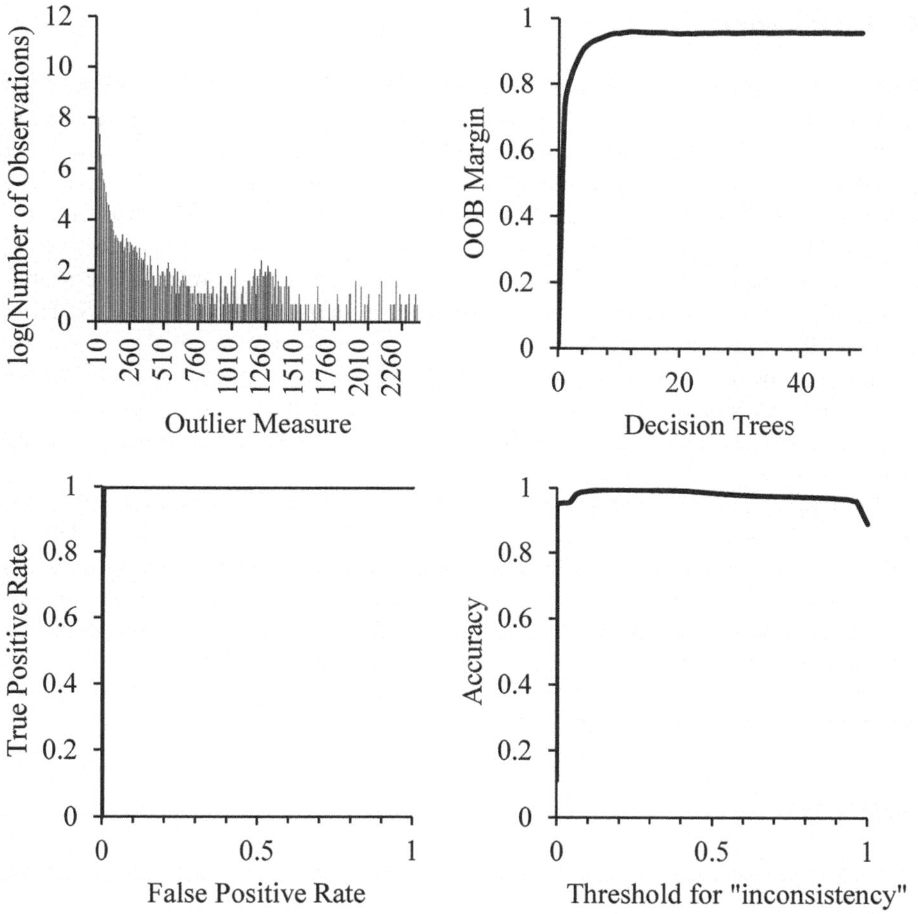

Fig. 7. Reduced amount of features, top left: outlier measurements, top right: out of bag margin, down left: receiver operating characteristics, down right: threshold of votes for class "inconsistency"

By comparing the two classifiers in Tables 3 and 5, it can be seen that the reduction of the features brings a slight improvement. The false positive rate is lowered by 0.1%, the true negative rate increases by 0.1%. As a result, specificity increases by 0.2% and sensitivity by 0.1%. This results in a total classification result improved by 0.1% overall with 88.7% correctly classified feature vectors. Likewise, the misclassifications mainly occur in measurements that are at the transition of classes. At the same time, both classifiers also have problems classifying the noisy measurements correctly.

Table 5. Confusion matrix reduced amount of features

	EM inconsistent	EM consistent	
OS inconsistent	40.6%	3.9%	91.3%
OS consistent	7.4%	48.1%	86.6%
	84.6%	92.6%	88.7%

4 Conclusion

Within this paper, an observer structure for monitoring the state of static environment modeling in a motor vehicle was derived and tested within a simulation environment. Features and the method random forest with bootstrap aggregation were derived for classification of the state. The classifier consists of 50 trained decision trees. The individual decision-trees are created based on a sample size that is obtained from the training data record using the method sample with replacement. There is a distinction between the classes "consistent" and "inconsistent." The classes refer to the usability of environment modeling for driver assistance systems respectively ADAS. Then, the classifier was optimized for its number of features, number of decision trees, and majority decision threshold parameters. The OS can distinguish applicable from clearly unusable situations of environmental modeling. In the area of transitions, however, there are misclassifications, with the rate of 8.7% fatal errors (false negative), which overlook an inconsistency.

In future work, the amount of data, especially inconsistent data, needs to be enlarged. To counteract the unlikely event of a malfunction, we propose to deliberately bring the prototype into situations where the environment modeling does not work properly. That can be done for example by a manual faulty calibration or artificial irrigation of the sensors. Furthermore, a better classification rate could be achieved by an extension to other parts of the EM, e.g., dynamic EM.

Acknowledgement. The authors gratefully acknowledge the support of Continental.

References

1. Statistisches Bundesamt. (ed.): Unfallentwicklung auf deutschen Strassen 2015, Berlin (2016)
2. Schlüter, D. Heterogene Sensordatenfusion zur robusten Objektverfolgung im automobilen Straßenverkehr. Carl von Ossietzky-Universität, Oldenburg (2003)
3. Maurer, M., Stiller, C. (eds.): Fahrerassistenzsysteme mit maschineller Wahrnehmung, p. 246. Springer, Berlin (2005)
4. Luenberger, D.G.: Observing the state of a linear system. IEEE Trans. Mil. Electron. **8**(2), 74–80 (1964)
5. Venkatasubramanian, V., Rengaswamy, R., Kavuri, S.N., Yin, K.: A review of process fault detection and diagnosis. Comput. Chem. Eng. **27**(3), 327–346 (2003)
6. Rasmussen, J., Vicente, K.J.: Coping with human errors through system design. Int. J. Man Mach. Stud. **31**(5), 517–534 (1989)
7. Kleer, J., Brown, J.S.: A qualitative physics based on confluences. In: Readings in Qualitative Reasoning About Physical Systems, pp. 88–126. Elsevier (1990)
8. Lin, C.-L., Liu, C.-T.: Failure detection and adaptive compensation for fault tolerable flight control systems. IEEE Trans. Industr. Inf. **3**(4), 322–331 (2007)
9. Jun, L., Ping, Z.: Sensor fault detection algorithm of flight control system under modeling uncertainty and noise disturbance, pp. 1851–1855 (2014)
10. Silipo, R., Adae, I., Winters, P.: Anomaly detection in predictive maintenance. In: Data Analysis, Machine Learning and Applications, 1. Aufl., S. 1. Springer (2008)

11. Elfes, A.: Using occupancy grids for mobile robot perception and navigation. Computer **22**(6), 46–57 (1989)
12. Shafer, G.: A Mathematical Theory of Evidence, p. 297. Princeton University Press, Princeton (1976)
13. Khoshgoftaar, T.M., van Hulse, J., Napolitano, A.: Comparing boosting and bagging techniques with noisy and imbalanced data. IEEE Trans. Syst. Man Cyber. Part A Syst. Hum. **41**(3), 552–568 (2011)
14. Ho, T. K.: Random Decision Forests, pp. 278–282. IEEE Computer Society Press, New York (1995)
15. Zhou, Z.-H.: Ensemble Methods, p. 234. CRC Press, Hoboken (2012)
16. Schapire, R.E.: The strength of weak learnability. Mach. Learn. **5**(2), 197–227 (1990)
17. Breiman, L.: Bagging predictors. Mach. Learn. **24**(2), 123–140 (1996)
18. Chawla, N.V., Lazarevic, A., Hall, L.O., Bowyer, K.W.: SMOTEBoost. In: Knowledge Discovery in Databases: PKDD 2003, vol. 2838, pp. 107–119. Springer, Berlin (2003)
19. Hido, S., Kashima, H. Takahashi, Y.: Roughly balanced bagging for imbalanced data. Stat. Anal. Data Min. **2**(5), 412–426 (2009)
20. Lichmann, M.: UCI Machine Learning Repository (2013). http://archive.ics.uci.edu/ml
21. Bija, A., Stuetzle, W.: Observations on bagging. In: Statistica Sinica, 16th edn., pp. 323–351. Institute of Statistical Science, Academia Sinica (2006)

Security in Hybrid Vehicular Communication Based on ITS G5, LTE-V, and Mobile Edge Computing

Jan-Felix van Dam[1(✉)], Norbert Bißmeyer[1], Christian Zimmermann[2], and Kurt Eckert[2]

[1] ESCRYPT GmbH, Bochum, Germany
jan-felix.vandam@escrypt.com
[2] Robert Bosch GmbH, Renningen, Germany

1 Introduction

The introduction of a Cooperative Intelligent Transportation System (C-ITS) is an enabler for increased safety on roads and future autonomous vehicle deployments. Further, traffic efficiency is intended to be increased by having smoother and more efficient traffic flow. New applications are planned in the area of driver convenience, public transportation and commercial carriage of goods. In this context, the term vehicle-to-everything (V2X) communication is used to describe real-time communication in transportation.

V2X communication is based on wireless radio technology that allows the transmission of short messages between enabled devices. Instead of using only one radio technology (ITS-G5), the 5G Automotive Association (5GAA) proposes to use cellular radio technology in addition to leverage V2X communications and to address future ITS use cases [1]. Cellular-V2X (C-V2X) radio technology considers the implementation in different broadband cellular network generations, i.e. LTE-V in 4G and future generations such as 5G.

In Sect. 2, the architecture of the considered C-ITS is described including the entities and communication paths. In Sect. 3, the radio-level communication technologies are introduced. The security requirements are summarized in Sect. 4 which are considered by security measures in Sect. 5. A conclusion is provided in Sect. 6.

2 Architecture of C-ITS

Several entities and systems are involved in communication in C-ITS. In this paper, we focus on a hybrid architecture that considers direct communication via ITS-G5 and C-V2X as well as indirect communication via cell tower. In Fig. 1, the considered communication paths between C-ITS stations are depicted.

T. Bertram (Hrsg.): *Fahrerassistenzsysteme 2018*, Proceedings, S. 80–91, 2019.
https://doi.org/10.1007/978-3-658-23751-6_8

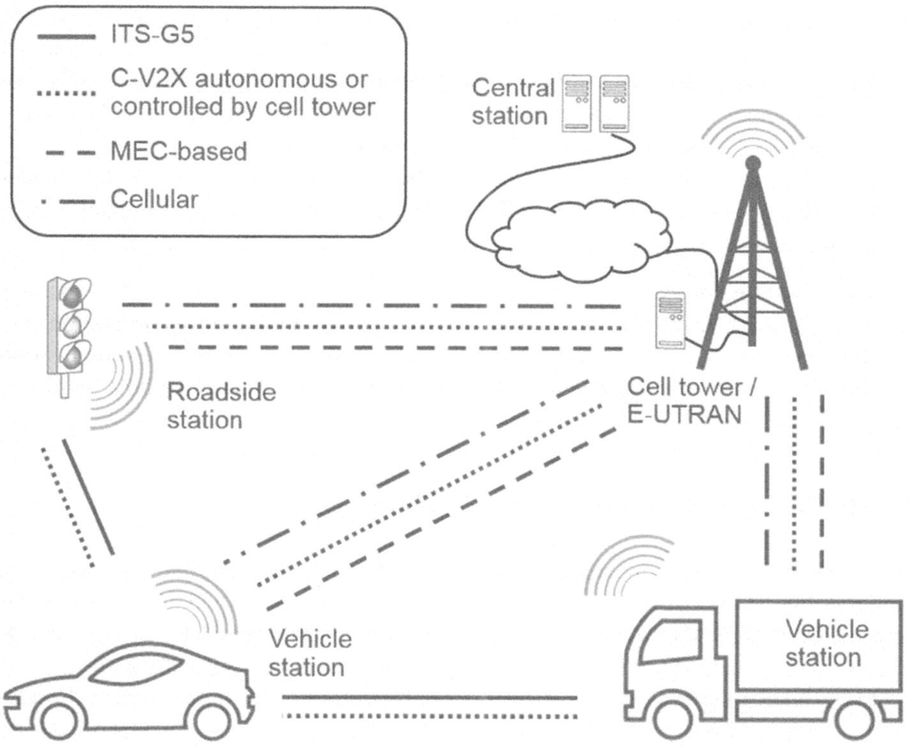

Fig. 1. Hybrid V2X communications architecture

- The end entities in the communication are referred to as C-ITS stations, which could be in practice any kind of vehicles, roadside units, two-wheelers or pedestrians.
- The required accurate time and location information is provided by Global Navigation Satellite System (GNSS).
- The cell tower may be involved depending on the C-V2X mode.
  - In "operator managed" mode, the cell tower provides network related actions to optimize direct V2X communication by supporting with channel utilization and quality of service. The Evolved-Universal Terrestrial Radio Access Network (E-UTRAN) is responsible for scheduling and interference management as well as synchronization and discovery. This supporting communication is performed via control signaling over the interface between C-ITS station and E-UTRAN. In the managed mode, the Public Land Mobile Network (PLMN) may also be responsible to perform access control to direct C-V2X autonomous communication by performing authorization tasks and transmitting permission parameters [2].
  - In classical cellular mode, the cell tower provides a data interface to the Internet in order to allow connecting with backend services, a C-ITS central station or a cloud service. Depending on the use case the station might only use the uplink or only the downlink to transmit or receive data. In particular, the cellular network might be used to multicast V2X messages to multiple receivers via Multimedia Broadcast Multicast Service (MBMS) [3].

  - In a Mobile Edge Computing (MEC) scenario, the cell tower hosts a V2X service to allow local communication not routed through backbone and internet to consider scalability and low latency requirements of the V2X use cases. For example, a V2X application running on the cell tower equipment may be used to relay event-driven V2X messages in order to cover a larger area.
- A C-ITS central station may be involved in the V2X communication as data provider or consumer. For example, a central Traffic Management Center collects traffic density information in order to calculate traffic light phases or detour information.

In case of autonomous V2X communication (ITS-G5 or C-V2X in "non-operator managed" mode), infrastructure components are not required for initialization and enabling end-to-end data exchange between C-ITS stations. For direct communication in C-V2X "operator managed" mode, the cell tower is involved to support the direct data exchange between C-ITS stations.

3 Radio-Level Communication Technologies

In a hybrid V2X communication scenario different radio technologies are used. As shown in Fig. 1, the C-ITS station may communicate with a backend using the cellular network via E-UTRAN, Mobility Management Entity (MME) and S/P-Gateway. In this case the cellular radio-level technology is applied as well as the related frequencies.

More interesting is the case where the C-ITS station communicates directly with another station in range. For this case the radio frequencies in the 5.9 GHz band are foreseen as depicted in Fig. 2. In this band, three safety channels (G5CC, G5SC1, G5SC2) can be used for ITS road safety and traffic efficiency applications according to ETSI ES 202 663 [4]. Since there is no direct limitation with respect to radio technology, a cellular radio (C-V2X) could use them in the same way as an ITS-G5 radio.

Even both radio-level communication technologies could coexist as long as different channels are used. One could imagine that for example the channel G5SC1 with 10 MHz is used by ITS-G5, the second channel G5SC2 with 10 MHz is shared by both radios and the third channel G5CC with 10 MHz is used by C-V2X [5]. However, it should be ensured that different technologies do not interfere each other, for example by out of band emissions, and that bandwidth is not wasted if same messages are broadcasted over multiple radios at the same time.

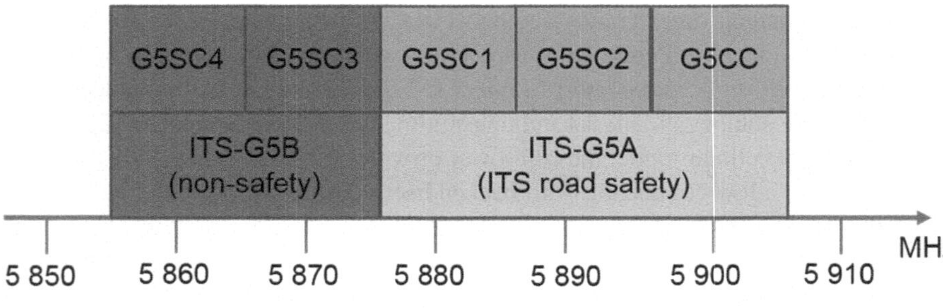

Fig. 2. C-ITS frequencies according to [4]

As a result, it might be reasonable to operate only one radio on the channels depicted in Fig. 2.

In the following list, the four different communication channels are described that are shown in Fig. 1.

- Device-to-device communication
  - via ITS-G5 based on IEEE 802.11p [6],
  - via C-V2X based on 3GPP Release 14 [2] (PC5 reference point for autonomous "non-operator managed" and "operator managed" device-to-device communication).
- Device-to-cell tower communication
  - using C-V2X based on 3GPP Release 14 [2] (V3 reference point via LTE-Uu to optimize the direct "operator managed" device–to–device communi-cation by scheduling and other tasks),
  - via cell tower using MEC (LTE-Uu reference point for device-to-cell tower communication).

Device-to-network communication using traditional cellular links is based on LTE (LTE-Uu reference point via network gateway to provide device-to-Internet communication according to 3GPP [3]).

4 Security Requirements

ITS communication is threatened due to its open character. Each participant of road traffic including vehicles, bicycles, pedestrians, etc. should be able to send and receive trustworthy information concerning traffic safety and efficiency. According to the ETSI threat, vulnerability and risk analysis [7], the most critical threats are the denial of transmission and reception of data, modification and deletion of transmitted information, mas-querade of a station, and acquisition of personal information.

For all use cases, data integrity, sender authentication and authorization, replay protection, and availability is required whereas confidentiality and accountability is only required for selected use cases.

In addition, system security of C-ITS stations and cellular network equipment is identified in the 3GPP report on V2X security [3] as relevant security requirement. In the following list, the C-ITS security requirements are summarized.

- Message integrity has to be ensured in order to detect manipulations made by attackers. System and software integrity of C-ITS station equipment has to be ensured in order to protect keys, certificates, and configurations against manipulation. The secure environment of the station may check the integrity of the entity's boot process.
- Authenticity and authorization of senders shall be ensured to allow only trusted systems to participate actively in C-ITS communications. The provision of authentication and authorization shall be revocable to exclude systems from C-ITS if necessary. Further, the secure environment of stations shall be protected against illegitimate access.
- Replay of valid messages has to be detected.

- Availability of C-ITS services should not be prevented by malicious activity.
- Confidentiality of V2X message content is required if a C-ITS service shall only be used by selected authorized stations. Sensitive data such as symmetric and private keys in transition and at rest shall not be exposed to external entities. Credentials for cellular network access shall reside on the Universal Integrated-Circuit Card (UICC).
- Privacy protection is required in order to prevent ITS-internal or external entities to collect personal identifying information. Neither a network operator nor a security infrastructure operator shall be able to link a pseudonymous identifier of a C-ITS station to its long-term ID. Furthermore, it shall not be possible to link different pseu-donyms to the same owner. Attackers shall not be able to track a vehicle by collecting V2X messages from the wireless radio.
- Detection of malicious station behavior shall be considered in order to achieve trust and confidence in V2X messages on top of cryptographic authentication measures.

In this paper we focus on security in hybrid C-ITS. Non-security requirements such as low latency, throughput and costs need to be considered but are not further analyzed in detail.

5 Security Concept for Hybrid V2X Communication

The security concept provides security measures for all requirements listed in Sect. 4. An overview of existing solutions is given in Sect. 5.1 and detailed with focus on the security goals in Sect. 5.2.

5.1 Overview of Existing Security Concepts

There are at least two relevant security architectures for C-ITS that have been developed in Europe [8] and the USA [9]. The European variant is standardized by European Tele-communications Standards Institute (ETSI) [10] and is the basis for the European C-ITS Certificate Policy (C-ITS CP) [11]. In contrast, the US variant is the basis for initial deployments under control of the US Department of Transportation.

Both solutions rely on the same basic principle. Each C-ITS station owns at least one asymmetric key pair whereby the public key is part of a certificate, which is signed by a trusted Certificate Authority (CA) that is part of a Public Key Infrastructure (PKI).

For securing a V2X message, the sender is signing the message payload with its private key and transmits the signed message together with its certificate. Each receiver is then able to trust in the integrity of the message by verifying the message signature and is able to trust in the authentication of the originator by verifying the signature of the sender's certificate.

Since this general security proceeding is harmonized in the European and US variants, the formats for security header and certificate are based on the same IEEE standard [12]. In particular, the ETSI security formats [13] are described as a profile of the

IEEE 1609.2 standard [12]. A V2X specialized certificate format is used to minimize the size of secured messages which contain the certificate of the sender in the security header.

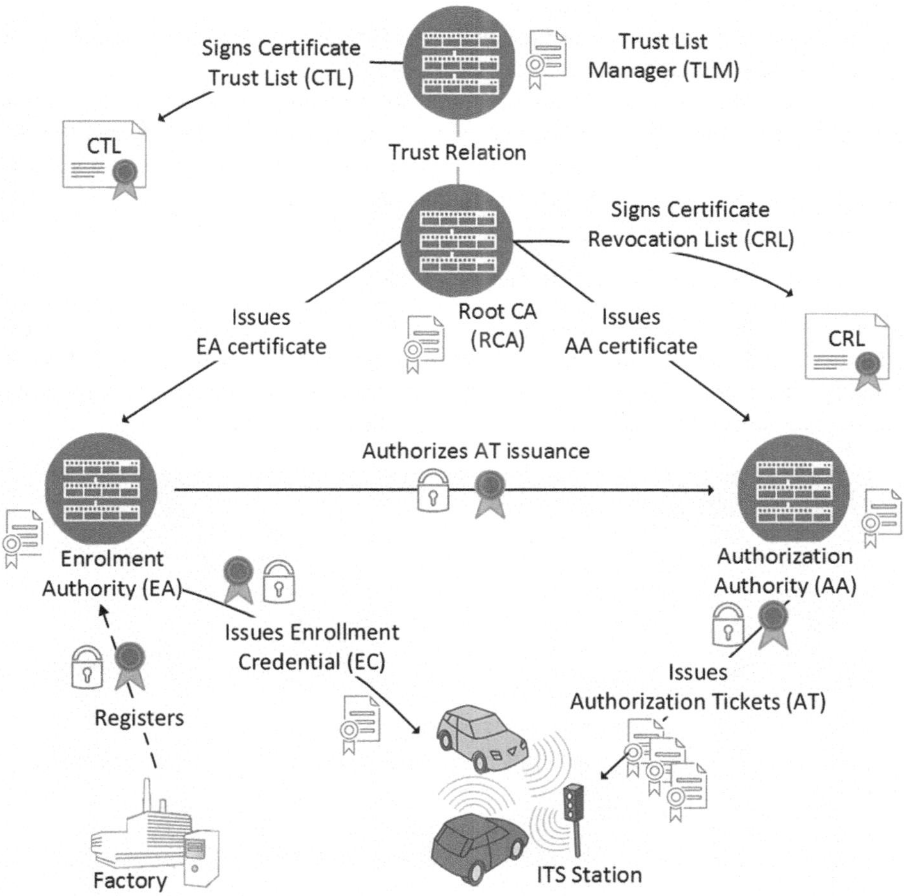

Fig. 3. C-ITS public key infrastructure [11]

Main difference between both security architectures can be found in the structure and functionality of the PKI. In this paper we focus on the European PKI variant shown in Fig. 3.

The Root CA (RCA) is the trust anchor in the system. Since there might be several RCAs in a C-ITS, a Trust List Manager (TLM) creates and distributes a signed Certificate Trust List (CTL) containing root certificates of trusted RCAs as well as access information to contact the RCAs from the end entities.

Each C-ITS station needs to be registered at one Enrolment Authority (EA) in order to get a valid pseudonym certificate for the V2X communication in form of an Authorization Ticket (AT). For private C-ITS stations the EA and AA need to be operated by different organizations in order to protect the station's privacy. The AA gets the request for a new AT but is not able to identify the requesters identity because the signer of the AT request can only be decrypted by the EA. After issuing an AT certificate the EA only knows that a specific trusted station has requested an AT but has never seen any content of the AT. On the other hand, the AA has issued the AT without having seen the identity of the requester.

The message protection in cellular networks is mainly based on a symmetric key architecture described in [14].

As a result, there exists a security solution for protecting the direct V2X communication between the stations (PC5 link) and for the communication between the station and the cell tower (LTE-Uu link). In the following section we propose a security concept considering the hybrid V2X communication scenario.

5.2 Security Concept Protecting Security Goals

According to the security requirements listed in Sect. 4, message integrity, sender authenticity and authorization, replay protection, and privacy shall be ensured on all channels whenever a V2X message is transmitted. The communication stack used in a hybrid communication system could look like as illustrated in Fig. 4.

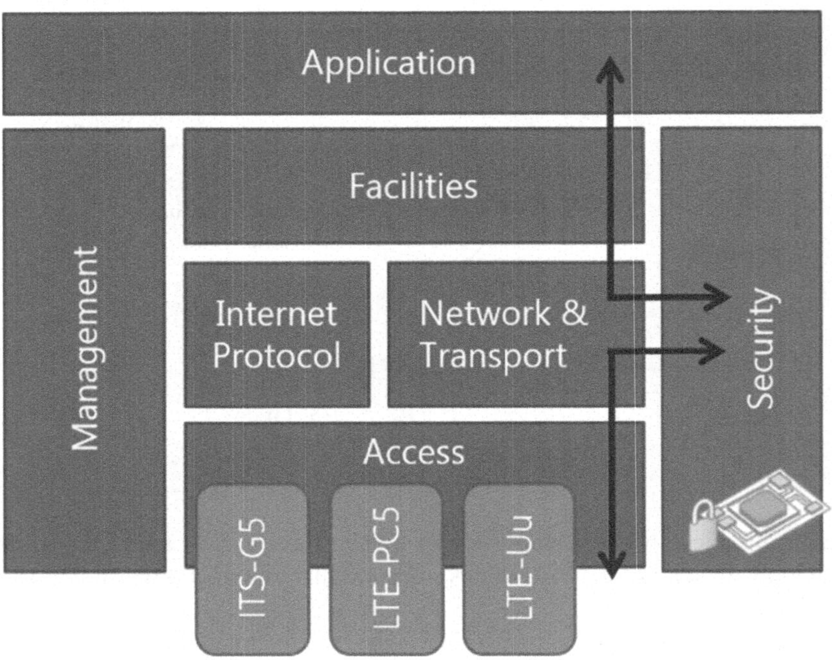

Fig. 4. Hybrid V2X communication stack (European C-ITS variant)

On the sender C-ITS station, the V2X message is generated on application or facilities layer and then handed over to the network & transport layer. On this layer the interface of the security component is used to include the security header into the V2X packet. A reference specification of this interface is given by ETSI [15].

Depending on the information provided by the upper layer, i.e. required sender permissions, security operation (sign/sign & encrypt), etc., the security component checks that an appropriate pseudonym certificate is loaded that can be used to sign the payload. The security layer creates a security header according to ETSI [13] including a signature over the payload, signer information in form of a digest or an AT certificate, generation time, and the payload Application ID (AID). If the confidentiality of the message payload needs to be protected, the signed payload is encrypted in a second step with a symmetric key. Information about the applied symmetric key are added to the security header to allow one or multiple receivers to select the correct key for decryption. Since the encrypted message is also used to protect the privacy of the communicating end entities the message is first signed and then encrypted.

By applying security measures on network & transport layer the security goals can be fulfilled independent of the radio-level communication technology.

This security solution can also be applied to secure known hybrid communication scenarios. In the device-to-device communication over ITS-G5 or C-V2X, the C-ITS station and the cell tower based roadside unit applies the ETSI ITS security [13] to secure V2X messages on network & transport layer [16]. In the device-to-cell tower communication, the end entities apply LTE security according to [14] to establish a mutual authenticated security channel based on symmetric cryptography. V2X messages sent through this channel are secured in addition with the AT certificates according to [13]. That enables end-to-end security if data is transmitted from a vehicle to an application server in the cloud where the LTE network is covering only a part of the communication. In this scenario, the application server would like to verify that the V2X message was generated by a valid and trustworthy C-ITS station. In a MEC scenario where the cell tower distributes V2X messages to neighboring cells the receiving station needs to check integrity and freshness of the original message and authenticity and authorization of the originator or the message.

Message Integrity. Message integrity is ensured by creating a digital signature over the message payload and packet routing information using the private key related to AT certificate that is currently loaded by the sending station. The AT key pair needs to be generated, administered and stored inside a cryptographic security hardware module [11].

Sender Authentication. Sender authentication is given by adding the AT certificate to the security header. The receiver needs to be equipped with all root certificates which are distributed as CTL. The signature of the AT certificate can be verified with the public key contained in the certificate of the AA and the validity of the AA certificate can be checked with the public key contained in the securely stored root certificate. In case an AA certificate is not available, the receiver can request the missing certificate from the sender on demand using the ETSI protocol [13].

Only trusted and certified C-ITS stations should be able to get a valid AT certificate from the PKI. In a first step, the station or the communication device is registered at the EA with a non-standardized process. Subsequently, the C-ITS station generates an asymmetric key pair for an Enrolment Credential (EC) and sends an EC certificate request to the EA using the ETSI protocol [17]. The EA issues the EC certificate with a potentially long validity lifetime and replies the certificate back to the station.

The EA has to ensure that only registered devices are able to request an EC and that requests from blacklisted or deactivated stations are rejected. In order to get an AT certificate from an AA, the C-ITS station is generating an asymmetric key pair, puts the public key into an AT request message according to [17] and signs it with the EC private key. The AA issues the AT certificate and provides it back to the requesting station.

A receiver of a V2X message can trust the owner of an AT as long as the related root in the certificate chain is trusted. The cryptographic algorithms ensure that only one private key can belong to an AT certificate that cannot be guessed by an attacker.

Sender Authorization. The authorization of the sender is considered by having encoded the sender's permission in the AT certificate in form of AIDs. Also the permissions of the AA and root can be limited by having AIDs also in their related certificates.

Replay Detection. Replay detection is enabled by adding a generation timestamp to the security header. Depending on the use case, a receiver can reject messages older than some seconds. For messages with longer lifetimes, the ITS station stores the hash digest of previously received messages. Incoming messages can then be compared with that list of previously received messages.

Confidentiality. Confidentiality can be ensured by encrypting the signed packet with a key shared with the recipient.

Revocation of Trust. Revocation of misbehaving or compromised C-ITS stations is enabled by using AT certificates with a short lifetime. The AA of the PKI checks that only AT certificates are issued with a relatively short validity period. In addition, AT certificates are only issued with validity for the near future. The specific parameters for private and non-private roadside stations are specified in the C-ITS CP [11].

Revocation of root certificates is done by removing the affected certificate from the CTL. An AA certificate, however, is revoked with a Certificate Revocation List (CRL) which is issued by the RCA that issued the AA certificate. As a consequence, the receiver needs to ensure that the latest CTL and CRL is used when the certificate chain from AT to root certificate is checked.

Privacy Protection. Privacy protection is considered by measures performed by the message sending C-ITS station, the cellular network, and the PKI.

A mobile C-ITS station, i.e. vehicle, needs to ensure that it is equipped with multiple AT certificates that are valid for the same period of time. Depending on the time and moved distance, the station changes all identifiers contained in a V2X message. In particular, the IDs added to the packet by the different layers of the communication stack need to be changed in a coordinated way together with the AT certificate. Even if most V2X messages are not encrypted, an attacker listing on the wireless channel cannot link a message to a specific person/vehicle since the V2X message does not contain identifying information. In addition, a local attacker is not able to link different pseudonyms of a station to each other since the attacker can only receive messages inside its communication range which is usually limited to a few hundred meters.

The PKI is designed in a way that no single entity is able to link a pseudonym of a station to its real or long-term identity, cf. Sect. 4. To prevent that an AA can collect a list of all pseudonyms of a station, the C-ITS station changes its IP address between each request of AT certificate according to [17].

In order to protect the privacy against the Mobile Network Operator (MNO), the station can hide its real identity (International Mobile Subscriber Identity, IMSI) and change its pseudonym identity (P-IMSI) frequently by re-attaching to the E-UTRAN. As proposed by 3GPP [3], a dedicated V2X Mobile Virtual Network Operator (MVNO) can be introduced to hide the IMSI against the MNO. Only the vehicle and the Home Subscriber Server (HSS) of the V2X MVNO know the IMSI. The MNO is handling only a pseudonymous IMSI in order to perform required tasks. If a linking of IMSI and P-IMSI is required, for example for legal investigations, the MNO and MVNO have to cooperate.

For the communication between a C-ITS station and an application server in the cloud using a cellular network, end-to-end encryption with changing IDs needs to be applied [3]. For the purpose of securing the hybrid V2X communication and for protecting the driver's privacy, the ETSI ITS security formats and procedures could be applied [5, 13].

Reliability. Ensuring the reliability of communication capabilities and channel bandwidth is challenging since an attacker can jam the channel with random noise. As discussed in Sect. 3, different frequency bands could be used in parallel. However, by sending the same messages on neighboring channels, valuable frequency resources would be wasted and an attacker could jam the entire 5.9 GHz band which is reserved for C-ITS.

For V2X communication without low latency requirements, messages could be transmitted via device-to-device communication using ITS-G5 or LTE-PC5 and in parallel via device-to-cell tower communication using LTE-Uu. In the latter case, the cell tower could apply the enhanced Multimedia Broadcast Multicast Service (eMBMS) to distribute V2X messages efficiently.

6 Conclusion

Securing V2X communication is essential in order to prevent attacks and ensure trust in received V2X data. For a hybrid V2X solution with different radio-level communication technologies and different modes, all security requirements need to be considered.

The security solution as specified by ETSI [10] and IEEE [12] for ITS-G5 communication can be applied without major modifications. The C-ITS station applies security measures for message protection on network & transport layer independent of the radio technology used on the access layer. The standardized PKI can be used to issue pseudonymous certificates for the ITS-G5 and C-V2X communication.

The hybrid vehicular communication has the potential of making direct and low latency data exchange more reliable without introducing additional risks from security point of view.

Acknowledgements. This paper is created in context of the project CONCORDA which is co-financed by the European Union's Connecting Europe Facility. The sole responsibility of this publication lies with the author. The European Union is not responsible for any use that may be made of the information contained therein.

CONCORDA

Co-financed by the European Union
Connecting Europe Facility

References

1. 5G Automotive Association: The case for cellular v2x for safety and cooperative driving. (2016). Online
2. ETSI – European Telecommunications Standards Institute: TS 123 285 v14.4.0 – Universal Mobile Telecommunications System (UMTS); LTE; Architecture enhancements for V2X services (3GPP TS 23.285 version 14.4.0 Release 14). ETSI, 3GPP, LTE (2017)
3. 3rd Generation Partnership Project: TR 33.885 – 3rd Generation Partnership Project; Technical Specification Group Services and System Aspects; Study on security aspects for LTE support of Vehicle-to-Everything (V2X) services (Release 14). 4GPP (2017)
4. ETSI – European Telecommunications Standards Institute: ES 202 663 v1.1.0 – Intelligent Transport Systems (ITS); European Profile Standard for the Physical and Medium Access Control Layer of Intelligent Transport Systems Operating in the 5 GHz Frequency Band. ETSI (2010)
5. Qualcomm: Accelerating C-V2X Commercialization. Qualcomm (2017)
6. IEEE Computer Society: IEEE Std 802.11p – Telecommunications and Information Exchange between Systems – Local and Metropolitan Area Networks – Specific Requirements – Part II: Wireless LAN Medium Access Control (MAC) and Physical Layer (PHY) Specific. IEEE (2010)
7. ETSI – European Telecommunications Standards Institute: TR 102 893 v1.2.1 – Intelligent Transport Systems (ITS); Security; Threat, Vulnerability and Risk Analysis (TVRA). ETSI (2017)
8. Bissmeyer, N., Stotz, J.P., Stübing, H., Schoch, E., Götz, S., Lonc, B.: A generic public key infrastructure for securing Car-to-X communication. In: 18th World Congress on Intelligent Transportation Systems (2011)
9. Whyte, W., Weimerskirch, A., Kumar, V., Hehn, T.: A security credential management system for V2V communications. In: IEEE Vehicular Networking Conference (VNC) (2013)
10. ETSI – European Telecommunications Standards Institute: TS 102 940 v1.1.1 – Intelligent Transport Systems (ITS); Security; ITS Communications Security Architecture and Security Management. ETSI (2012)
11. C-ITS Platform: Certificate Policy for Deployment and Operation of European Cooperative Intelligent Transport Systems (C-ITS). European Commission (2017)
12. IEEE Computer Society: IEEE P1609.2a-2017 – IEEE Standard for Wireless Access in Vehicular Environments – Security Services for Applications and Management Messages. Institute of Electrical and Electronics Engineers (2017)

13. ETSI – European Telecommunications Standards Institute: TS 103 097 v1.3.1 – Intelligent Transport Systems (ITS); Security; Security Header and Certificate Formats. ETSI (2017)
14. ETSI – European Telecommunications Standards Institute: TS 133 401 V14.4.0 – Digital cellular telecommunications system (Phase 2+) (GSM); Universal Mobile Telecommunications System (UMTS); LTE; 3GPP System Architecture Evolution (SAE); Security architecture (3GPP TS 33.401 version 14.4.0 Release 14). ETSI (2017)
15. ETSI – European Telecommunications Standards Institute: TS 102 723-8 – Intelligent Transport Systems (ITS); OSI Cross-Layer Topics; Part 8: Interface between Security Entity and Network and Transport Layers. ETSI (2013)
16. ETSI – European Telecommunications Standards Institute: EN 302 636-4-1 v1.3.1 – Intelligent Transport Systems (ITS); Vehicular Communications; Geonetworking; Part 4: Geographical Addressing and Forwarding for Point-to-Point and Point-to-Multipoint Communications; Sub-part 1: Media-Independent Functionality. ETSI (2017)
17. ETSI – European Telecommunications Standards Institute: DRAFT TS 102 941 – Intelligent Transport Systems (ITS); Security; Trust and Privacy Management. ETSI (2018)

Automated Driving – Misunderstandings About Level 3 and Euro NCAP's Activities in this Field

Andre Seeck(✉), Elisabeth Shi, and André Wiggerich

BASt, Bergisch Gladbach, Germany
seeck@bast.de

1 Fundamentals and Misunderstandings

This section aims to resolve common misunderstandings about Level 3 automated driving. It will commence with a brief overview over the fundamentals of automated driving, focusing to distinguish between Level 2 and Level 3. Next, common confusions about Level 3 automation will be addressed and put straight.

1.1 Fundamentals of Automated Driving

In order to describe the full picture of vehicle automation, [1] proposes a classification scheme based on three Principles of Operation taking action on the layer of vehicle guidance (cp. Fig. 1) [2]. These Principles of Operation constitute a superordinate structure and can be subdivided as follows:

Principle of Operation A summarizes informing and warning functions that can be subdivided into status information, abstract warnings and concrete warnings. These functions take indirect effect on vehicle guidance since only driver action can realize an effect. Principle of Operation B describes sustained automation that influences vehicle guidance directly. This Principle of Operation covers SAE Levels 1 to 5 (cp. Fig. 2). The Levels are described in detail in SAE-International J3016 [3]. Principle of Operation C classifies temporarily intervening functions with regard to their extent of operation and the type of hazard, in which they take action. For further details on the Principles of Operation, please refer to [1].

Differentiating Between SAE Level 2 and Level 3. Before addressing the misunderstandings concerning SAE Level 3, the level itself shall be described briefly and distinguished from the currently available Level 2 automation:

Currently available driving automation systems of SAE Level 2 cannot and do not perform the entire dynamic driving task. The driver determines whether and when to engage or disengage the system. After engagement, part of the dynamic driving task is performed by the driving automation system, i.e. lateral and longitudinal control. The remainder dynamic driving task, i.e. object and event detection and response, is performed by the driver. Furthermore, the driver supervises the driving automation system and intervenes if necessary.

T. Bertram (Hrsg.): *Fahrerassistenzsysteme 2018*, Proceedings, S. 92–97, 2019.
https://doi.org/10.1007/978-3-658-23751-6_9

Principle of Operation A: informing and warning	Principle of Operation B: sustained automation	Principle of Operation C: temporarily intervening in accident-prone situations
Only indirect influence on vehicle guidance via the driver. 1. Status information 2. Warning (abstract hazard) 3. Warning (concrete hazard)	Take direct influence on vehicle guidance (conscious activation by the driver – divided responsibilities in execution of the dynamic driving task). Always overrideable.	Preventive machine intervention in case of negative situation prediction. Either: I. driver as controller does not react conform to expectation <u>or</u> is inaccessible II. in accident-prone situations drivers/ controllers cannot handle due to performance limitations

Fig. 1. Principles of operation [1]

LEVEL	NAME	EXAMPLES	Dynamic driving task (DDT)[1]		[DDT] fallback[2]	ODD (Operational design domain)[3]
			Sustained lateral and longitudinal vehicle motion control	Object and event detection & response (OEDR)[4]		
Driver performs part or all of the dynamic driving task (DDT)						
0	No driving automation	warnings / active safety, only (if anything)	driver	driver	driver	unlimited (none)
1	Driver assistance	adaptive cruise control OR lane centering[5]	driver and system	driver	driver	limited
2	Partial driving automation	adaptive cruise control AND lane centering	system	driver	driver	limited
ADS[6] ("system") performs all of the dynamic driving task (DDT)						
3	Conditional driving automation	automated driving in freeway traffic	system	system	fallback-ready user[7]	limited
4	High driving automation	automated driving in geo-fenced locations	system	system	system	limited
5	Full driving automation	automated driving everywhere (taxi)	system	system	system	unlimited

Fig. 2. SAE-International Standard J3016 [4]

Quite the opposite is true for driving automation systems of SAE Level 3: They can and do perform the entire dynamic driving task within their operational design domain. The driving automation system only permits engagement of the system within its operational design domain. After the driver engages the driving automation system there (and the system accepts engagement), the driver becomes the fallback-ready user and therefore, does not perform the dynamic driving task anymore. However, as a fallback-ready user s/he needs to be prepared to take-over the dynamic driving task after the system announces a request to intervene.

The major difference between Level 2 and Level 3 regarding the role of the driver lies in the fact that whilst in Level 2 the driver still needs to permanently perform part of the dynamic driving him-/herself after having engaged the driving automation system, the driver of a Level 3 System does not drive him-/herself anymore after engaging, but becomes a "fallback-ready user". This amounts to an up to date completely new and unknown role of a "driver": With automation engaged the driving-related role is merely "awareness" in the sense of standing by to react to takeover requests. And only this reduction far-reaching automation can allow the execution of non-driving related tasks e.g. watching television.

1.2 Misunderstandings About Level 3

Some common misunderstandings will be addressed and resolved in the following

Misunderstanding 1: "Many legal issues still need to be solved". In discussions about market introduction of Level 3 systems it is often argued that many legal issues still need to be solved to allow manufacturers to offer certain driving automation systems in near future.

From an objective point of view, legal requirements can be divided into three aspects:

Road Traffic Act. To permit operation of Level 3 vehicles amendment of national traffic law is required.

Since the latest law amendment of the German road traffic act in June 2017, the operation of vehicle automation on SAE Level 3 or higher is legally permitted in Germany. Thus, this legal aspect has already been solved in Germany. Only the issues of technical approval at UN-level are still awaiting amendment (cp. below).

Liability/Insurance. In discussions on liability and insurance of Level 3 vehicles a gap in liability for accidents caused by automation is asserted. In this context a higher risk for policyholders constitutes further concern.

In Germany insurances cover both fault based and strict liability. Accidents caused by driving automation systems can be regarded as part of each vehicles "operational hazard" and are thus already covered by liabilities that are not fault-based (so called strict liability). Hence, from an objective point of view, the aspect of liability and insurance of Level 3 vehicles is not an issue in Germany.

Furthermore, in the light of product liability defective control by a driving automation system constitutes a product defect in case of Level 3. Therefore, manufacturers can be made accountable (e.g. by insurances) for accidents occurring during highly automated driving due to defective control by a driving automation system. Compared to Level 2, the clearly defined operational design domain of Level 3 systems facilitates the inference of a product defect substantially.

Technical (type) Approval. Another remark relates to the lack of regulations for technical (type) approval of Level 3 systems.

Regulations concerning technical (type) approval are addressed at international level. Currently, the UN is discussing regulations for Level 2 systems and has just begun discussions on the regulation of Level 3 driving automation systems.

In conclusion, the process of integrating vehicle automation on a regular basis into technical regulations and type approval has already started.

Misunderstanding 2: "At Level 3, take-over needs to be accomplished within 4 seconds." In discussions on Level 3 automation systems, it is frequently, yet erroneously, referred to UN-R-79, which in its current draft prescribes a time frame of at least 4 s for transition.

UN-R-79 applies to Level 2 automation systems only. Level 3 automation systems are out of scope (cp. Sect. Differentiating SAE Level 2 and Level 3). By definition at Level 2, the driver constantly monitors the driving environment as well as the driving automation system and intervenes if necessary. The frequently cited time frame of 4 s mentioned in UN-R79 applies to this context.

Time frames for take-over on Level 3 have not been addressed in a legal context, yet. These discussions on Level 3 automation systems have just begun (cp. Sect. *Technical (type) Approval*).

2 Euro NCAP's Activities

Even if highly automated driving functions (SAE Level 3 and higher) will roll out in new vehicles not until near future, semi-automated systems and assisted driving functions (SAE Level 2), i.e. Traffic Jam Assist or high speed Motorway Assist, have recently been introduced to the market. Since the role of the driver for these particular systems is to monitor not only the roadway but also the system permanently and react instantaneously to any upcoming potentially critical situation (e.g. objects suddenly appearing in the lane ahead, other vehicles cutting in at short distance, defect of a sensor etc.), the driver still remains the ultimate and immediate fallback level. This imposes great responsibility on the driver and fundamentally alters drivers' habitual strategies of driving (this is at least true for all drivers unacquainted to SAE Level 1 Driver Assistance Systems so far). Besides well established legal requirements (cp. Sect. *Misunderstanding 1: "Many legal issues still need to be solved."*), safety of an automated driving function therefore refers to and directly results from a broad and fundamental understanding of the intended use and limitations by the users. Consequently the consumer perspective taken by Euro NCAP plays an essential role for the introduction of safe automated driving.

As already reported, automated driving technologies can be understood to likewise cover temporarily intervening functions in near accident situations (crash avoidance technologies, e.g. AEB) on the one hand as well as sustained automation (a.k.a. automated driving functions; functions that can perform specific driving tasks, either fully, or temporarily and in particular circumstances) on the other hand. The Euro NCAP automated driving assessment will primarily focus on sustained automation (Principle of Operation B) with regard to different levels of automation, since crash avoidance technologies (Principle of Operation C) are already integrated in the active safety assessment protocols. As automated driving in terms of safety is likely to be seen and understood differently by consumers compared to active and/or passive safety, Euro NCAP's grading as well as the communication will be performed without compromising the current well established five-star rating system. Therefore a separate grading system for automated driving functions is needed.

Developing such a grading system purposefully, it is important to consider the general impact of automated driving on road safety. One key aspect that will contribute to road safety is the avoidance of critical situations based on human error. However, since the use of automated driving functions is still (and will also be in near future) limited to certain amount of use cases (e.g. motorway driving or parking), the by far greater benefit is to be expected to result from higher fitment rates of temporarily intervening functions in accident-prone situations (Principle of Operation C): They can be "always on", even outside the specific and limited domain of automated driving and, due to higher quantity and cost reductions, will also become available in multiple vehicles equipped with only lower levels of sustained automation (SAE Level 1–2) in future.

With respect to the impact described, Euro NCAP's objective is therefore to establish a high standard of safety for the market introduction of sustained automation and at the same time inform consumers independently on safe use. This seems to be one major aspect since reliable and easily comprehensible sources of information are still rare and tend to disregard consumers' capacities. In addition, advertising and marketing in a world of digitisation have a tendency to exaggerate.

To realize the important goal of safe automated driving, Euro NCAP has already implemented a Strategic Group on Automated Driving for the grading and communication. Two subordinate working groups are developing methods and procedures for standardized vehicle assessments in terms of Test Procedures and Human-Machine-Interaction. Industry is involved in both working groups. The Test Procedures group is developing a test matrix for longitudinal und lateral control tests considering stationary and moving targets as well as cut-in and cut-out scenarios in order to map technical performance. The Human-Machine-Interaction group is planning to perform a checklist-based expert assessment for first introduction and will come up with a holistic assessment also including representative user assessment samples in addition on the long run. The objective for this group is to take not only the commonly discussed human machine interface into account but to focus strongly on interactional processes regarding the user's understanding of his role and responsibility when applying automated driving functions as well as the user's awareness of the particular automation modes or traffic situations.

Performing a phased-in approach in the second half of 2018, Euro NCAP is planning to issue a single publication on all available systems at that point in time. This publication will inform consumers on the optional driver assistance systems available (SAE Level 2). Information will be based on human-machine-interaction assessment as well as physical testing to demonstrate the limitations of these systems. Protocols for higher levels of automation (SAE Level 3 and higher) will be prepared by the working groups in near future. Euro NCAP's general approach implies different grading scales per technology covering continuous assistance on the one hand as well as basic, advanced and superior automation on the other hand to describe and assess the differences between redundant-parallel driving (assistance; SAE Level 2) and independent machine driving (automation, SAE Level 3 and higher). To allow a distinction between the specifications of different systems and functions, the grading scales will be use-case or domain specific (e.g. parking, highway, traffic jam, etc.) while at the same time taking different levels of automation into account (technology by technology).

In general, the aim of Euro NCAP will be the safe introduction of automated driving to the market striving for safe use by consumers for all kinds of different assisted and automated driving functions. Taking into account that only assisting functions (SAE Level 2) are presently available on the market, too often wrong information help lull drivers into a false sense of safety when applying these systems and thus giving rise to exaggerated expectations. The truth for actual systems is that drivers still need to be constantly vigilant and attentive at every moment when behind the wheel. Helping consumers by getting this message across will therefore be crucial.

References

1. Gasser, T.M., Frey, A.T., Seeck, A., Auerswald, R.: Comprehensive definitions for automated driving and ADAS. In: 25th International Technical Conference on the Enhanced Safety of Vehicles (ESV) National Highway Traffic Safety Administration (2017)
2. Donges, E.: Driver behavior models. In: Winner, H., et al. (ed.) Handbook of Driver Assistance Systems (Basic Information, Components and Systems for Active Safety and Comfort), vol. 1 (2016)
3. SAE International Standard J3016: Taxonomy and definitions for terms related to on-road-motor vehicle automated driving systems (2016)

Putting People Center Stage – To Drive and to be Driven

Klas Ihme, Katharina Preuk, Uwe Drewitz, and Meike Jipp(✉)

DLR, Braunschweig, Germany
meike.jipp@dlr.de

Technological advancements allow the implementation of systems that can assist humans and even perform tasks that were previously carried out manually. To classify the multiple ways with which such systems can provide services to humans, generic frameworks were introduced (e.g., Jipp and Ackerman 2016; Kaber and Endsley 1997; Parasuraman et al. 2000). For example, Parasuraman et al. (2000) based their framework on a generic perception-cognition-action cycle and used four functions in which systems can assist humans. These functions include information acquisition, information analysis, decision making, and action implementation. If a system executes one of these functions more autonomously, its level of automation with regard to this specific function increases. If, for instance, a night vision system detects a pedestrian and warns a driver, this system certainly assists in information acquisition and information analysis. Yet, it does not assist with regard to decision making and action implementation. This and related frameworks are independent from the area of application. Although this has advantages for establishing and evaluating theories about human reactions to automated systems, such frameworks can be criticized because they leave room for interpretation with regard to the specific application. This is why application-specific levels of automation were introduced (e.g., BAST 2012; Brandenburger and Jipp 2017). In the automotive domain, the BAST (2012) differentiated

- Driver-Only Mode, in which drivers continuously execute the lateral and transverse driving task.
- Assisted Mode, in which drivers continuously execute either the lateral or the transverse driving task. The system executes the remaining task. Yet, drivers have to monitor the systems' actions and react in case it becomes necessary.
- Partially-Automated Mode, in which the automation continuously executes the lateral and transverse driving task in specific situations. Drivers have to monitor the automation and be ready to take over the tasks in case it becomes necessary.
- Highly-Automated Mode, in which the automation continuously executes the lateral and transverse driving task in specific situations. Drivers do not have to monitor the automation, because it recognizes its boundaries, and asks its driver to take over the driving task within a certain time frame.

T. Bertram (Hrsg.): *Fahrerassistenzsysteme 2018,* Proceedings, S. 98–108, 2019.
https://doi.org/10.1007/978-3-658-23751-6_10

– Fully-Automated Mode, in which the automation executes the lateral and transverse driving task in specific situations. Here, the automation is able to bring the system into a state of minimal risk. Consequently, drivers do no longer need to react in case the automation reaches its boundaries.

An analysis of these driving modes reveals two challenges highly relevant for the introduction of automated system: interaction with human road users and driver states. The interaction with human road users is relevant because automated cars are integrated into a complex environment of human traffic participants who expect certain behaviour of other traffic participants. A wrong expectation may lead to a negative change of cognitive and emotional states (e.g., emotions, workload). Such negative states can in turn cause human reactions eventually leading to safety-critical situations. Within the scope of this paper, it will thus be analysed first how human drivers react when they encounter automated cars. Subsequently, it will be analysed how human states can be interpreted automatically—a method highly relevant for the implementation of empathic automation. Empathic automation is a concept with which automated systems can cooperate with humans such that negative consequences can be avoided.

1 Interaction with Human Road Users

Automated cars can have information about how driving situations will develop because of their use of various camera systems, radar and laser systems, car-to-car-, and car-to-infrastructure communication (Bartels et al. 2011; Fleming 2012). For instance, an automated car can make use of a traffic light assistance system. This system enables more time- and energy-efficient driving at signalised intersections. It uses car-to-infrastructure communication and is herewith informed about traffic light signal switches. Based on this information, a highly-automated system can efficiently approach traffic lights (Kettwich et al. 2016). In case the system uses an assisted mode, it can provide drivers with either speed or action recommendations. It can, for instance, recommend its drivers to slow down in the approach to a green traffic light to avoid unnecessary strong acceleration and deceleration. Such anticipatory driving behaviour has the potential to increase traffic flow at signalized intersections and reduce environmental impacts (Draskoczy et al. 1998; Frey et al. 2003; Rakha et al. 2000; Staubach et al. 2014).

The penetration rate of automated cars will initially be small (Proskawetz et al. 2013). One consequence may be that the expected positive effects of such technology on criteria such as traffic throughput at an urban intersection can hardly be reached. Consequently, researchers used traffic simulations to identify the effects of increasing penetration rates on relevant criteria (Arem et al. 2006; Katsaros et al. 2011; Krajzewicz et al. 2012). In these studies, human drivers' behaviour was based on driver models that have constraints in terms of integrating human factors, such as emotions (Krajzewicz and Wagner 2004; Olstam and Tapani 2004). For example, Katsaros et al. (2011) evaluated the benefit of a traffic light assistance system with different penetration rates. The driving behaviour of human drivers in car-following scenarios was based on the assumption that they adapted their driving behaviour to preceding cars with traffic light assistance system only when certain thresholds in distances were reached. However, human driving behaviour is rather complex and amongst other factors, shaped by expectations (e.g., Alexander and Lunenfeld 1979; Brookhuis et al. 2001; Martens 2000, 2004; Martens and Fox 2007; Näätänen and Summala 1976; Olson and Sivak 1986). Human drivers may not expect this kind of anticipatory driving behaviour (Preuk 2018) and thus adapt their behaviour not adequately resulting in safety-critical situations.

Expectations of Human Drivers When They Encounter Automated Cars. In a recent study, we forced human drivers to encounter an automated car without prior knowledge to do so (Preuk et al. 2016a, b). Specifically, three drivers drove in a multi-driver simulator, in which multiple participants can complete the experiment simultaneously as they drive together in the same virtual environment. This concept of a multi-driver simulator had been realized in the MoSAIC (Modular Scalable Applications Platform for ITS Components, for a detailed description see Lorenz et al. 2011; Oeltze and Schießl 2015). Such an experimental infrastructure has various advantages. First, instead of making use of driver models, participants can encounter each other. Thus, the interaction of drivers can be studied more validly in a multi-driver simulation (Mühlbacher 2013). Second, encounters of participants can be realized repeatedly, and safety-critical situations can also be targeted (Mühlbacher et al. 2018; Preuk et al. 2018).

In this multi-driver simulator study, three participants drove in a car-following scenario in an urban area with signalized intersections. The lead driver was a confederate followed by two human drivers. The confederate's vehicle was equipped with a specific function of a traffic light assistance system–the so-called *start-up assistance system* that was developed by Volkswagen within the project UR:BAN (Poppe and Kranke, personal communication, April 11, 2013). With this system, the confederate came to a halt several meters away from the stop line at a red traffic light. The confederate began accelerating before the traffic lights turned green and crossed the stop line with a higher speed at the moment the traffic light switched to green. Consequently, the time it usually takes to react to signal changes was reduced so that the traffic flow at signalized intersections could be increased. Two naïve human drivers followed this confederate.

Thirty experimental trials were executed, and data of 60 drivers (40 male, 20 female) between the ages of 20 and 78 years (*mean* = 30.6 ± 10.8 years) were collected.

The study followed a mixed design: One factor manipulated within participants was the confederate's equipment with the two levels "with" or "without" the start-up assistance system. The participants experienced the scenario with and without the confederate's equipment in randomized order. The other factor manipulated within participants was the number of intersections as the human drivers repeatedly drove through signalised intersections. Another factor yet manipulated between participants was the parametrisation (two levels) of the start-up assistance system: In the moderate parametrisation, the assisted confederate came to a halt 4 m from the stop line and started up 1.3 s before traffic signal change, whereas in the extreme parametrisation, the assisted confederate came to a halt 10 m from the stop line and started up 2.6 s before traffic signal change.

The data analyses revealed that human drivers benefited from the driving behaviour of the confederate with start-up assistance system: They adapted their driving behaviour significantly and crossed the stop line earlier after the signal change as well (Preuk et al. 2016a). However, such encounters also resulted partly in safety issues and negative ratings: The following human drivers had smaller headways when the confederate came to a halt 10 m from the stop line indicating late or less sufficient brake reactions (Preuk et al. 2016b). Moreover, human drivers rated the automated driving behaviour as aversive and did not understand its benefit (Preuk et al. 2016a, b).

Implications: The Relevance of User States. The negative ratings reported by Preuk et al. (2016a, b) can be an indicator for human drivers feeling frustrated by automated cars' driving behaviour. Frustration occurs when goal directed behaviour such as driving is blocked (Lazarus 1991). The early stop of the confederate may have been perceived as blocking the own way. This conforms to the results of Frehse (2015), who found drivers became angry when they encountered a preceding car to brake unexpectedly.

Frustrated drivers are more likely to engage in aggressive, safety-critical driving behaviour (Deffenbacher et al. 2002; Mesken et al. 2007; Shinar and Compton 2004). This can not only explain the occurrence of the smaller headways in the previous study but also has a negative influence on the acceptance of passengers of automated cars (cf. Carsten and Fowkes 1998; Malone et al. 2014; Saad et al. 1999). User states are thus a crucial topic for the introduction of automated cars.

2 User States

Even though people have excellent skills to drive cars, their driving abilities are prone to be negatively influenced by various factors. Physiological, emotional and cognitive states such as fatigue, frustration and increased workload can have serious effects on peoples' driving skills and can cause serious impairments, some of which drivers even may not be aware of. As highlighted before, negative emotions such as frustration are associated with aggressive driving behavior and higher risks to be involved in accidents (Mesken et al. 2007; Shinar and Compton 2004). Similar holds true for increased mental workload, which can lead to inattention for critical events in the surrounding traffic or for relevant aspects of the environment and which therefor is a main factor for the occurrence of accidents and dangerous situations (cf. Rumar 1990). Negative consequences related to user states can be avoided, if technology is available that can diagnose the current state of its user, interpret it and react appropriately. In the following, it will be described how technology can be enabled to diagnose the current state of its user.

2.1 Mental Workload

The violation of expectations, for example, a car that behaves in a way that deviates clearly from our experiences in the past, causes surprise. Surprise, in turn, triggers a variety of cognitive processes. According to Meyer et al. (1997), the appraisal of an event as unexpected produces an interruption of ongoing information processing and the reallocation of mental resources. These resources are needed for the analysis and evaluation of the unexpected event. This reallocation is necessary to ensure an up-to-date knowledge base and accurate predictions for future actions. However, if mental resources are reallocated when performing a dynamic task in a dynamic environment, this can cause serious issues. The concept of mental work-load (cf. Wickens 2008) reflects this utilization of the driver's cognitive resources. According to this concept, increasing mental workload limits the attention that can be payed to additional new information and can have a negative impact on the driver's awareness of the current driving situation (cf. Endsley 1995). Especially when available mental resources of drivers who are processing the current traffic situation or analysing unexpected events are used to full capacity, an increase in demands cannot be adequately compensated for, leading to delayed actions, e.g., late brake reactions. For this reason the design and development of driver assistance as well as car automation should take workload information into account, either offline as an input to a human-centered design process or online to adapt systems and human-machine interaction to individual demands.

Mental workload (i.e., the extent to which cognitive resources are used by the tasks being processed) is reflected in physiological measures such as heart rate (Wilson and Eggemeier 1991), heart rate variability (Backs and Seljos 1994; Mulder and Mulder 1981), respiration (Veltman and Gaillard 1998; Wientjes 1992), the skin resistance (Helander 1978; Lindholm and Cheatham 1983), and the pupil size (Beatty 1982; Kahnemann and Beatty 1966). However, beside the generally given sensitivity of these measures for mental workload, it has to be evaluated whether the correlations between an increase of workload and the corresponding change of these measures as described in the literature can be found under real-world driving conditions too.

The assessment of mental workload using various physiological methods was shown in a simulator study (Unni et al. 2017). In this study, the 15 participants drove on a three-lane highway and had to accomplish a working memory task (N-back task) with five different levels of difficulty. The N-Back task was integrated into the speed regulation: The participants had to drive the speed that was presented on the speed sign N presentations behind. This means that in case of a 0-back task, the speed that was displayed on the last passed sign had to be driven. In a 1-Back-Task, the speed of the penultimate shield had to be driven, and so on. In the study, N was varied from 0 to 4. During the drives, both, cardiovascular measurements (heart rate, heart rate variability) and neurophysiology were measured using functional near infrared spectroscopy (fNIRS). Then, multivariate regression and 10-fold cross-validation were used to estimate the extent to which physiological data could be used to predict the mental workload. The data analyses showed first, that a prediction of different levels of mental workload can be accomplished based on cardiovascular measurements. Second, the fNIRS measurements showed that brain areas associated in previous studies with the updating and retention of working-memory contents were increasingly activated in the processing of brain areas (Lara and Wallis 2015; Postle 2006). Thus, the model determined by the physiological parameters could be checked for plausibility by the neurophysiological measures. Overall, the study confirmed that mental workload can be measured on the basis of physiological parameters even in realistic scenarios.

2.2 Emotions

Emotions have a strong influence on behaviour and cognition (Scherer 2009). They impact cognitive processes relevant for driving including sufficient situation awareness (Jeon 2012, 2015). Moreover, humans are often not aware of the negative effects of emotions on driving and therefore, unlike in the case of distraction, do not attempt to compensate for it (Jeon 2015). This leads to reduced safety, if the human experiences strong negative emotions. One possibility to take human emotional needs into account is the development of user-focused systems that are capable of recognising the emotional states of the vehicle occupants, derive their current needs in order to offer specific assistance that balances the emotional state or counteracts maladaptive behaviours. Importantly, developing such systems requires a valid and robust assessment of emotions. Up to now, no fully automated systems to measure emotions of drivers exist.

Emotions can be considered as collective, synchronised changes of the organic sub-systems of the human that occur in reaction to events classified as relevant for the organism (Scherer 2005, 2009). A measurement of emotions should therefore be based on these changes in the subsystems. Mostly, five subsystems are discriminated: A cognitive evaluation (appraisal) of the event takes place in the central nervous system (CNS). This triggers a physiological adaptation serving to regulate the organism which is generally reflected in an alteration of the arousal level. Moreover, motivational changes are elicited leading to changes in behavior. In addition, emotions are communicated by changes in the voice, mimics, and posture. Finally, monitoring all the aforementioned changes in the CNS leads to a change in emotional experience (feeling) (cf. Scherer 2005). In order to gain an exact image of the current emotional state of a person, as many of these components as possible should be considered. Nonetheless, for real-time applications in the car, sensors should be unobtrusive for the vehicle occupants. Therefore, camera- or microphone-based methods, sensors for measuring physiology integrated into the vehicle interior or evaluating the driving behavior seems to be promising approaches. In the following, the results of a research project will be presented that indicates the potential of measuring driver frustration using video recordings of the face (Ihme et al. in press, 2017).

Frustration was induced through a combination of time pressure and blocking goal attainment in a driving simulator study. For this, participants (31 in total, 12 female, age: mean = 26.2 ± 3.5 years) were promised an extra amount of money if they made to drive through an urban scenario in a specific amount of time. The drive was blocked multiple times by several obstacles, such as red traffic lights or slow lead vehicles, which led to a significant increase in frustration. In a control condition, the participants could drive without blocking events. During the drives, the participants' faces were videotaped, so that the facial expressions during frustrated driving could be investigated. In a first step, one-minute segments of each condition were manually coded according to the Facial Action Coding System (Ekman et al. 2002). The analyses showed that frustrated drivers activated muscles in the mouth area (lip tightening and lip pressing) more frequently than non-frustrated ones (Ihme et al. in press). In the next step, it was evaluated whether these results could be replicated with an automated analysis of the video data. For this, the participants' facial muscle activity was extracted from each video frame by means of a commercial software package (Emotient Facet, Imotions, Copenhagen). Then, patterns of facial muscle activity were determined by means of a cluster analysis and frequency of occurrence of these was compared between the conditions. Again, the activity of muscles in participants' mouth region changed: A combination from lip pucking, lip pressing, and chin raising was shown more often in the frustrated than in the control condition (Ihme et al. 2017). Interestingly, the revealed patterns are in line with earlier studies on frustration from the domain of human-computer interaction (Grafsgaard et al. 2013; D'Mello et al. 2005).

Altogether these results confirm the potential of video-based assessment of facial expression to infer the emotional state of the driver.

2.3 Implications: The Potential for Empathic Automation

The available results about user states show the potential for the assessment of workload and emotions using video-based assessment of facial expressions and physiological changes. Such information about current levels of workload and emotions can be used to parametrise user-focused assistance functionalities preventing inadequate human behavior to influence traffic safety negatively. Such systems can then be considered empathic automation systems.

3 Conclusions: Empathic Automation

Humans exhibit and interpret a variety of verbal and nonverbal cues when they communicate with each other. Examples for nonverbal cues are gestures, facial expressions, eye movements, speech melody and prosody, and physiological reactions, like blushing or sweating. By interpreting the context, a listener can understand the other's physiological and emotional state and anticipate the goals, intentions, beliefs and actions of her/his counterpart which gives her/him the possibility to act and react adequately. The competence underlying this ability to understand and react accordingly is empathy.

Empathic automation thus first needs to be enabled to perceive cues from different information sources of the human counterpart, merge this information and integrate contextual factors (environment, situation and task) as well as domain-specific knowledge, to be able to understand the human cognitive and emotional state, and anticipate future behaviour. For this purpose, empathic automation certainly needs reliable, preferably unobtrusive sensor systems, sophisticated machine-learning algorithms being able to correctly interpret the data deriving from the sensor systems, and actor systems with which it can influence cognitive and emotional states and future human behaviour. Then, the automation has knowledge about the current state of the user, about why this state developed, what its consequences are, and how it could be mitigated. Such systems will certainly not be equipped with empathy "tout court" (cf. Stephan 2015), but "only" with empathy that can handle relevant situations in the field of driving or being driven. Nevertheless, the implementation of empathic automation in the field of driving is a complex problem.

Within the scope of this paper, knowledge could be provided for the types of sensor systems required to assess the mental workload and emotional state of users. Specifically, in-vehicle camera systems are needed. These systems can provide data about physiological changes (i.e., heart-rate related measures) and about facial expressions (e.g., lip pressing, chin raising). Analyses algorithms can use these data and provide information about the development of the current cognitive and emotional state of the user. If future research is able to identify ways with which negative cognitive and emotional states can be mitigated automatically, empathic automation can harmonise the states of the user to enable adequate human reactions. One way to mitigate negative cognitive and emotional states may be to inform negatively impacted human drivers about the surrounding automated cars and their abilities (cf. Preuk 2018). Another way may be to address the emotional reaction of the user and assist drivers establishing a more positive emotional reaction. Whatever may be an adequate, empathic reaction to a negative cognitive and emotional state, it will be an important step to make automation empathic, so that it can actually cooperate with humans to achieve a common goal: reaching a certain position safely and efficiently.

Acknowledgments. The UR:BAN project initiative was supported by the German Federal Ministry of Economics and Technology. The authors would like to thank Dr. Holger Poppe and Florian Kranke of Volkswagen AG – Environment (Strategy and Mobility) for the cooperation in UR:BAN. The research about driver states was supported by the funding initiative Niedersächsisches Vorab of the Volkswagen Foundation and the Ministry of Science and Culture of Lower Saxony as a part of the Interdisciplinary Research Centre on Critical Systems Engineering for Socio-Technical Systems.

References

Alexander, G.J., Lunenfeld, H.: Positive guidance in traffic control. In: Proceedings of the Human Factors and Ergonomics Society 23rd Annual Meeting, Santa Monica, USA (1979)

Van Arem, B., Van Driel, C.J.G., Visser, R.: The impact of cooperative adaptive cruise control on traffic flow characteristics. IEEE Intell. Transp. Syst. **7**(4), 429–436 (2006)

BAST: Rechtsfolgen zunehmender Fahrzeugautomatisierung. Berichte der Bundesanstalt für Straßenwesen, 83 (2012)

Backs, R.W., Seljos, K.A.: Metabolic and cardiorespiratory measures of mental effort: the effects of level of difficulty in a working memory task. Int. J. Psychophysiol. **16**(1), 57–68 (1994)

Bartels, A., Karrenberg, S., Weiser, A.: Hochautomatisches Fahren auf der Autobahn/Highly automated driving on the motorway. Verkehrsunfall Und Fahrzeugtechnik **49**(11), 2–6 (2011)

Beatty, J.: Task-evoked pupillary responses, processing load, and the structure of processing resources. Psychol. Bull. **91**(2), 276 (1982)

Brandenburger, N., Jipp, M.: Effects of expertise for automatic train operations. Cogn. Technol. Work **19**, 699–709 (2017)

Brookhuis, K., de Waard, D., Janssen, W.: Behavioural impacts of advanced driver assistance systems–an overview. Eur. J. Transp. Infrastruct. Res. **1**(3), 245–253 (2001)

Carsten, O., Fowkes, M.: External Vehicle Speed Control: Phase I Results, Executive Summary. Institute for Transport Studies. University of Leeds, Leeds (1998)

Deffenbacher, J.L., Lynch, R.S., Oetting, E.R., Swaim, R.C.: The driving anger expression inventory: a measure of how people express their anger on the road. Behav. Res. Ther. **40**(6), 717–737 (2002)

D'Mello, S., Craig, S., Gholson, B., Franklin, S., Picard, R., Graesser, A.: Integrating affect sensors in an intelligent tutoring system. In: Affective Interactions: The Computer in the Affective Loop Workshop at 2005 International Conference on Intelligent User Interfaces. AMC Press, New York (2005)

Draskoczy, M., Carsten, O., Kulmala, R.: Road safety guidelines. CODE project, telematics application programme, deliverable B5.2 (1998). http://cordis.europa.eu/pub/telematics/docs/tap_transport/code_b5.2.pdf

Ekman, P., Friesen, W.V., Hager, J.: Salt Lake City: A Human Face (2002)

Endsley, M.R.: Toward a theory of situation awareness in dynamic systems. Hum. Factors **37**(1), 32–64 (1995)

Fleming, B.: New automotive electronics technologies. IEEE Veh. Technol. Mag. **7**, 4–12 (2012)

Frehse, B. (2015). Kognitive Bewertungsdimensionen von Ärger im Straßenverkehr. Doctoral dissertation, Leuphana Universität

Frey, H.C, et al.: On-road measurement of vehicle tailpipe emissions using a portable instrument. J. Air Waste Manage. Assoc. **53**(8), 992–1002 (2003)

Grafsgaard, J.F., Wiggins, J.B., Boyer, K.E., Wiebe, E.N., Lester, J.C.: (2013)

Houtenbos, M., Hagenzieker, M.P., Wieringa, P.a., Hale, A.R.: The role of expectations in interaction behaviour between car drivers. Traffic Transp. Psychol. Theory Appl. **1**, 303–314 (2005). http://repository.tudelft.nl/islandora/object/uuid:d8aeddfe-892e-4b95-9cce-7cbbdb3797cc/datastream/OBJ/view

Helander, M.: Applicability of drivers' electrodermal response to the design of the traffic environment. J. Appl. Psychol. **63**, 481–488 (1978)

Ihme, K., Dömeland, C., Freese, M., Jipp, M.: Frustration in the face of the driver: a simulator study on facial muscle activity during frustrated driving. Interac. Stud. (in press)

Ihme, K., Zhang, M., Jipp, M.: Automatische Erkennung der Frustration von Autofahrern: Ergebnisse und Anwendungsmöglichkeiten. In: Automatisierungssysteme, Assistenzsysteme und eingebettete Systeme für Transportmittel – AAET 2017 (2017)

Jeon, M.: A systematic approach to using music for mitigating affective effects on driving performance and safety. In: The 2012 ACM Conference on Ubiquitous Computing, pp. 1127–1132, ACM, New York (2012)

Jeon, M.: Towards affect-integrated driving behaviour research. Theor. Issues Ergon. Sci. **16**, 553–585 (2015)

Jipp, M., Ackerman, P.: The impact of higher levels of automation on performance and situation awareness: a function of information-processing ability and working-memory capacity. J. Cognitive Eng. Dec. Making **10**, 138–166 (2016)

Kaber, D., Endsley, M.: Out-of-the-loop performance problems and the use of intermediate levels of automation for improved control system functioning and safety. Process Safety Rep. **16**, 126–131 (1997)

Kahneman, D., Beatty, J.: Pupil diameter and loadon memory. Science **154**, 1583–1585 (1966)

Katsaros, K., Kernchen, R., Dianati, M., Rieck, D.: Performance study of a green light optimal speed advisory (GLOSA) application using an ontegrated cooperative ITS simulation platform. In: Proceedings of 7th Wireless Communications and Mobile Computing Conference (IWCMC), Istanbul, Turkey, pp. 918–923 (2011)

Kettwich, C., Haus, R., Temme, G., Schieben, A.: Validation of an interaction concept indicating the status of the traffic light signal in the context of automated driving in urban environment. In: IEEE Intelligent Vehicles Symposium, Göteborg, Schweden (2016)

Krajzewicz, D., Bieker, L., Erdmann, J.: Preparing simulative evaluation of the GLOSA application. In: Proceedings of ITS World Congress, Vienna, Austria (2012)

Krajzewicz, D., Wagner, P.: Ansätze zur kognitiven Simulation eines Autofahrers. MMI-Interaktiv **7**, 84–97 (2004)

Lara, A.H., Wallis, J.D.: The role of prefrontal cortex in working memory: a mini review. Front. Syst. Neurosci. **9**, 173 (2015)

Lazarus, R.S.: Progress on a cognitive-motivational-relational theory of emotion. Am. Psychol. **46**(8), 819–834 (1991)

Lindholm, E., Cheatham, C.M.: Autonomic activity and workload duringlearning of a simulated aircraft carrier landing task. Aviat. Space Environ. Med. **54**(5), 435–439 (1983)

Lorenz, T., Baumann, M., Jaschke, K., Köster, F.: A modular and scalable application platform for testing and evaluating ITS components. In: Proceedings of 20th International Workshops on Enabling Technologies: Infrastructure for Collaborative Enterprises, Paris, France (2011)

Malone, K., Rech, J., Hogema, J., Innamaa, S., Hausberger, S., Dippold, M., van Noort, M., de Feijter, E., Rämä, P., Aittoniemi, E., Benz, T., Burckert, A., Enigk, H., Giosan, I., Gotschol, C., Gustafsson, D., Heinig, H., Katsaros, K., Neef, D., Ojeda, L., Schindhelm, R., Sütterlin, C., Visintainer, F.: DRIVE C2X – deliverable D11.4: impact assessment and user perception of cooperative systems (2014). http://www.drive-c2x.eu/tl_files/publications/Deliverables%20and%20abstracts/DRIVE%20C2X_D11.4_Impact_Assessment_v1.0_full%20version.pdf

Martens, M. H.: Automatic visual information processing and expectations in traffic. KFB & VTI FORSKNING/RESEARCH 35A. Linköping, Sweden (2000)

Martens, M.H.: Stimuli fixation and manual response as a function of expectancies. Hum. Factors J. Hum. Factors Ergon. Soc. **46**(3), 410–423 (2004)

Martens, M.H., Fox, M.: Does road familiarity change eye fixations? A comparison between watching a video and real driving. Transp. Res. Part F **10**, 33–47 (2007)

Mesken, J., Hagenzieker, M.P., Rothengatter, T., de Waard, D.: Frequency, determinants, and consequences of different drivers' emotions: an on-the-road study using self-reports, (observed) behaviour, and physiology. Transp. Res. Part F Traffic Psychol. Behav. **10**(6), 458–475 (2007)

Meyer, W.-U., Schützwohl, A., Reisenzein, R.: Toward a process analysis of emotions: the case of surprise. Motiv. Emot. **21**, 251–274 (1997)

Mulder, G., Mulder, L.J.: Information processing and cardiovascular control. Psychophysiology **18**, 392–402 (1981)

Mühlbacher, D.: Die Pulksimulation als Methode zur Untersuchung verkehrspsychologischer Fragestellungen. Doctoral dissertation, Julius-Maximilians-Universität Würzburg (2013)

Mühlbacher, D., Preuk, K., Lehsing, C., Will, S., Dotzauer, M.: Multi-road user simulation: methodological considerations from study planning to data analysis. In: Bengler, K., Hoffmann, S., Manstetten, D., Neukum, A., Drüke, J. (Hrsg.) UR:BAN Human Factors in Traffic, pp. 403–418. Springer Verlag, Wiesbaden (2018)

Näätänen, R., Summala, H.: Road User Behavior and Traffic Accidents, North Holland, Amsterdam (1976)

Oeltze, K., Schießl, C.: Benefits and challenges of multi-driver simulator studies. IET Intell. Transp. Syst. **9**(6), 618–625 (2015)

Olson, P.L., Sivak, M.: Perception-response time to unexpected roadway hazards. Hum. Factors **28**(1), 91–96 (1986)

Olstam, J.J., Tapani, A.: Comparison of Car-Following Models. Swedish National Road and Transport Research Institute, Linköping, Sweden (2004)

Parasuraman, R., Sheridan, T.B., Wickens, C.D.: A model for types and levels of human interaction with automation. IEEE Trans. Syst Man Cybern. **30**, 286–297 (2000)

Postle, B.R.: Working memory as an emergent property of the mind and brain. Neuroscience **139**, 23–38 (2006)

Preuk, K.: Encountering automated cars – what are consequences for human drivers. Doctoral dissertation, Technische Universität Carolo-Wilhelmina zu Braunschweig (2018)

Preuk, K., Dotzauer, M., Jipp, M., Köster, F.: Encounters between drivers with and without cooperative intelligent transport systems. In: Bengler, K., Hoffmann, S., Manstetten, D., Neukum, A., Drüke, J. (Hrsg.) UR:BAN Human Factors in Traffic, pp. 363–377. Springer Verlag, Wiesbaden (2018)

Preuk, K., Stemmler, E., Jipp, M.: Does surrounding traffic benefit from an assisted driver with traffic light assistance system? Transp. Res. Part F Traffic Psychol. Behav. **43**, 302–314 (2016a)

Preuk, K., Stemmler, E., Schießl, C., Jipp, M.: Does assisted driving behavior lead to safety-critical encounters with unequipped vehicles' drivers. Accid. Anal. Prev. **95**, 149–156 (2016b)

Proskawetz, K.-O., Klug, S., Beckert, B.: Smart cities and communities – key to innovation integrated solution – cooperative intelligent transport systems and services (C-ITS) (2013). https://eu-smartcities.eu/sites/all/files/Cooperative%20Intelligent%20Transport%20Systems%20and%20Services%20-%20Smart%20Cities%20Stakeholder%20Platform%20january.pdf

Rakha, H., Aerde, M. Van, Ahn, K., Trani, A.A.: Requirements for evaluating traffic signal control impacts on energy and emissions based on instantaneous speed and acceleration measurements. Transp. Res. Rec. J. Transp. Res. Board **1738**, 56–67 (2000)

Rumar, K.: The basic driver error: Late detection. Ergonomics **33**, 1281–1290 (1990)

Saad, F., Mundutéguy, C., Darses, F.: Managing interactions beetween car drivers: an essential dimension of reliable driving. In: Proceedings of Seventh European Conference on Cognitive Science Approaches to Process Control (CSAPC1999) (1999)

Scherer, K.R.: What are emotions? And how can they be measured? Soc. Sci. Inf. **44**, 695–729 (2005)

Scherer, K.R.: Emotions are emergent processes: they require a dynamic computational architecture. Philos. Trans. R. Soc. B Biol. Sci. **364**, 3459–3474 (2009)

Shinar, D., Compton, R.: Aggressive driving: an observational study of driver, vehicle, and situational variables. Acc. Anal. Prev. **36**(3), 429–437 (2004)

Staubach, M., Schebitz, N., Köster, F., Kuck, D.: Evaluation of an eco-driving support system. Transp. Res. Part F Traffic Psychol. Behav. **27**, 11–21 (2014)

Stephan, A.: Empathy for Artificial Agents. Int. J. Soc. Robot. **7**(1), 111–116 (2015)

Unni, A., Ihme, K., Jipp, M., Rieger, J.W.: Assessing the driver's current level of working memory load with high density functional near-infrared spectroscopy: a realistic driving simulator study. Front. Hum. Neurosci. **11**, 167 (2017)

Veltman, J.A., Gaillard, A.W.K.: Physiological workload reactions to increasing levels of task difficulty. Ergonomics **41**(5), 656–669 (1998)

Wickens, C.D.: Multiple resources and mental workload. Hum. Factors **50**, 449–455 (2008)

Wientjes, C.J.: Respiration in psychophysiology: methods and applications. Biol. Psychol. **34**(2), 179–203 (1992)

Wilson, G.F., Eggemeier, F.T.: Psychophysiological assesment of workload in multitask environments. In: Damos, D.L. (Hrsg.) Multiple-Task Performance, S. 329–360. Taylor & Francis, London (1991)

Towards Positive User Experience (UX) for Automated Driving

Guido Meier-Arendt(✉)

Continental Automotive GmbH, Frankfurt, Germany
guido.meier-arendt@continental-corporation.com

1 Changing Human-Machine-Dialogue

It is a matter of time until vehicles will be capable of chauffeuring the driver when he/she prefers that. After all, vehicles will gradually take over more and more driving tasks over the oncoming years so that the driver turns into a user who is free to engage in a variety of other pursuits. Prior to that, however, the driver needs to learn trusting the new technology and to develop an awareness of the current driving situation, which in turn means that the elements of the human-machine interaction need to exert a positive influence on the user experience (UX). The prerequisite for a positive UX is a holistic way of taking the driver's preferences, requirements, the current situation, and the driver's environment into account.

Another key success factor of automated driving (AD) is that user and vehicle shall speak a common language because without this element the holistic human-machine interface cannot meet the UX criteria. Already today Continental is developing and critically scrutinizing suitable solutions. As a part of this process the company is carrying out in-depth testing in the driving simulator and on the road.

2 A Matter of Trust

In all likelihood the future drivers of automated vehicles will first have to overcome an inhibiting threshold of some individual form before they will delegate the driving task to the vehicle and begin to rely easily onto the automation. The interaction concept for this role reversal – and the subsequent back-delegation – needs to be ready for the widest imaginable spectrum of behavior patterns, ways of reacting, and driver expectations. This can extend as far intentional misuse. After all, it is conceivable that individual drivers may simply wait and see what will happen, if they do not respond to the request of taking the driving task back. A system maturity ready for this can hardly be reached in the lab because it is plain impossible to anticipate the full bandwidth of reaction patterns and speeds. A human-machine interface – consciously

G. Meier-Arendt is Principal Expert Human Machine Interface at Continental in Babenhausen (Germany).

T. Bertram (Hrsg.): *Fahrerassistenzsysteme 2018,* Proceedings, S. 109–114, 2019.
https://doi.org/10.1007/978-3-658-23751-6_11

referred to as "interaction concept" in this context – may be premeditated so well on a lab-scale, but at the end of the day it has to prove itself on the road.

In order to gain realistic insight into the UX quality, Continental has taken a new direction with the so-called UX research vehicle: In this car, a specially trained driver in the passenger seat takes over the longitudinal and lateral control task instead of an automation. This "AD driver" is curtained-off from the traditional driver workplace behind the wheel (Fig. 1). For the test person driving the vehicle, this AD driver in the passenger seat is officially a mere precaution in case the automation should reach a limit. For the test person the road trial feels just like exploring an automation on a public road while in fact the AD driver is acting the part of the automation at all times.

Fig. 1. The UX research vehicle

A test supervisor in the rear seat observes the audible and visible reaction patterns of the test driver and controls the process. During the driving, the test supervisor keeps asking the test person for his/her subjective impressions. The guiding research interest behind this is to find out how different test drivers react to the individual elements of the interaction established by sound/voice, image/symbol (icon), color/light, and vibration mainly during three phases: One is during delegating the driving task to the vehicle, number two is the automated driving mode itself and the initiation of driving maneuvers, and number three is delegating the driving task back to the test driver. It is of particular interest to see were uncertainties arise or rather how trust can be created effectively. This road-testing is complemented by simulating automated driving in the driving simulator of the ergonomics laboratory at Continental in Babenhausen (located to the Southeast of Frankfurt). In the lab, the eye movements and the quality of the driving task are monitored as well (Fig. 2).

Fig. 2. AD simulation complete with eye tracking carried out in the ergonomics lab

3 The User is Part of the System

During the first test driver road trials with a group of ten voluntary subjects unrelated to the topic, two aims were pursued: Firstly the above question whether the technologies of the interaction concept are suitable to build trust, and secondly the even more basic question whether the UX research vehicle is suitable to ascertain this.

The various communication modes (visual, acoustic, haptic) of the interaction concept employed during the roughly 30 min duration of the drive together establish an iterative concept which is designed to tell the driver clearly what the AD system is doing, what it is planning, and what the next step is. Top priority was to make transparent which actions are expected from the test driver. The AD system's output that shall achieve this consists of sounds (earcons), information in natural language, color-coded information (among others coming from an LED light strip), and ultimately, towards the end of an escalation plan, a vibrating seat. The test driver can enter his/her commands via a central control element which is reserved for automation purposes, such as maneuver initiation. Figure 3 shows the current system generation of this control element.

The sequence of natural-language information, distinctive earcons, and symbol-based presentation in the cluster instruments starts with first unobtrusive messages, accompanied by an earcon, followed by more emphatic signals and leads on to the tangible request to take back the wheel (seat vibration if the driver fails to react).

Fig. 3. The AD controller reserved for making AD-related entries

4 First Results Re. Transparency and Trust

The basis for evaluating the utilized interaction elements is provided by objective measurements (such as the time lapse between a takeover request of the automation and the moment when the driver actually puts his/her hands on the wheel) as well as so-called "thinking aloud". As part of this method the test supervisor repeatedly encourages the test driver to simply voice everything that comes to his/her mind while driving without filtering the thoughts. A kind of summary protocol of a typical test drive shall serve as a story line to give some first insights.

At the beginning of the "automated" driving section, the system tells the test driver that it is available and how to initiate the AD mode. This invitation is understood and after an initial hesitation, a rapid learning curve can be observed. The test drivers quickly feel safe and gain confidence.

During the "AD driving", the test driver is given the opportunity to choose from various display content qualities in the fully digital cluster instrument. Thus, the person behind the wheel can decide about the level of detail which is presented visually. A clear preference became obvious: The majority of the test drivers preferred the maximum detail level which depicts all detected vehicles around the ego-vehicle (Fig. 4). This much-preferred view of the environment shows what the automation sees and what it bases its control decisions on. Also it became clear that this detection should ideally be done in real time.

Fig. 4. Cluster instrument with the preferred view of the vehicle environment

Fascinating insight could be gained during the so-called Takeover Request (TOR), which is the request to the test driver to take back the driving task. At that moment of the test procedure, the drivers were intentionally involved in an activity not related to driving. Therefore, they had to finish this activity first. This resulted in a wide choice of reaction patterns. There were test drivers who immediately put their hands on the wheel, while other carefully scanned the environment first before touching the wheel, and there was a third group who stopped their ongoing activity, checked the environment and then waited for further instructions from the system. It was the latter reaction pattern that was actually intended: The test drivers were expected to mentally prepare by interrupting their ongoing activity but without any other immediate action. It became clear that the request for action to the test driver needs to be very specific and precise to create an awareness of the expected action without triggering any overly eager behavior. Independent from that nearly all drivers confirmed that they felt well-informed, with only a few saying differently. A similar 90/10 split has already been reported by other studies. For further interaction concept refinement, these outliers need to be analyzed in detail and suitable solution strategies need to be implemented as part of the dialogue.

From the combination of thinking aloud and subsequent qualitative interviews, which probed deeper into individual aspects, it became clear that while the preparation phase preceding the takeover worked well for many test drivers, there is still room for refinement. In particular, the process may benefit from an earlier start and a more finely graduated sequence. The haptic seat was also rated positive. The bandwidth of reaction suggests that it can be beneficial to make the interaction concept customizable so that, for instance, the tone of the earcons and the type of vibration can be adjusted.

During the "AD drive", the test supervisor gave the driver the task of initiating a lane change maneuver. This also resulted in insights, which are helpful to refine the activation sequence because the driver performance at the interface was actually better than expected. As a result, the sequence could be simplified. The test supervisor's request to set a higher driving speed also confirmed the usefulness of the UX vehicle. The set-speed function could be specifically honed after the first test driver study.

During the subsequent qualitative interviews the test drivers confirmed a high level of trust even though “AD driving” felt “still a little strange” as some put it. Among other things, future test driver trials will be longer and contain more demanding settings.

During this, the results from the first iteration will already be utilized to advance the maturity level of the interaction concept. As a bottom line it can be stated that the drivers are given time for other activities and that they appreciate this. They consider the instructions as very helpful, they realize that they need not do much themselves and that the maneuvers (lane change, set higher speed) work well.

5 Summary and Outlook

Using the UX research vehicle Continental has begun a user-centric design process, which serves to optimize the dialogue between driver and vehicle in a way that ensures a maximum level of transparency and trust. By integrating users at an early stage and through agile iteration cycles, it becomes possible to complement a mature automation solution with an equally mature interaction concept from day one. It will be through this combination alone that AD vehicles will gain credence and that they will be able to make a major contribution to a safer and more comfortable individual mobility. Even though the small population of the first test driver study does not permit any statistical evaluation, the UX research vehicle has proven itself as a very practical tool to validate which types of signal vehicles will employ to communicate in the future.

Systematically Generated and Complete Tests for Complex Driving Scenarios

Marc Habiger(✉), Marius Feilhauer, and Jürgen Häring

ETAS GmbH, Borsigstraße 14, 70469 Stuttgart, Germany
marc.habiger@etas.com

1 Introduction

A key success factor when realizing autonomous vehicles is the validation of their functionality. Due to their system architecture involving multiple environmental sensors, such as video cameras and LIDAR sensors, the input vector into the Advanced Driver Assistance System (ADAS) is high dimensional. The signal processing has to reliably execute the perception and cognition of the current driving situation [1]. The environment consists of an arbitrary number of elements, including traffic participants, material properties, weather conditions, road signs or buildings. Based on the availability of a semantic, machine-readable representation of scenarios, such driving situations can be described (see [2]). This allows the realization of a continuously growing test case database for the validation of autonomous driving functions (see [3, 4]).

In this publication we present a case study on how to automatically generate test cases using a model-based approach.

First we explain the fundamentals of closed-loop testing and mathematically describe our approach of executing tests of a system embedded in its environment. As the deviation of test cases requires the inversion of mathematical equations, we then present a practical approach of how to meaningful divide the infinite possibilities of concrete test values by using equivalence partitioning. We apply this approach to an exemplary use case and conclude this presentation by summarizing the approach.

Applying our approach, a large number of test cases can be generated automatically. The test effort is dramatically reduced by using the simulation of a virtual vehicle with ETAS virtualization tools.

ETAS provides a complete portfolio to realize efficient virtualization along all stages of the development cycle. ETAS COSYM allows to build a system simulation by connecting virtual ECUs (vECU) with simulation models. This tool is based on open standards to integrate 3rd party elements, as well as vECUs built with ETAS ISOLAR-EVE, and models from the ETAS LABCAR-MODEL library. Additionally, COSYM allows seamless transition between MiL, SiL and HiL setups, always providing the same access to the simulation (Fig. 1).

T. Bertram (Hrsg.): *Fahrerassistenzsysteme 2018,* Proceedings, S. 115–125, 2019.
https://doi.org/10.1007/978-3-658-23751-6_12

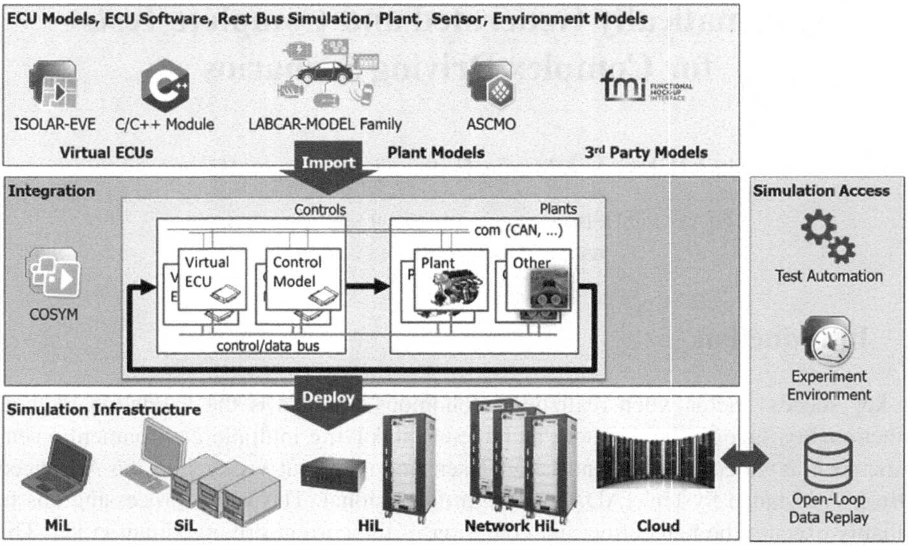

Fig. 1. ETAS virtualization solution

2 Closed-loop Testing Fundamentals

A Unit under Test (UuT) is described as a system f characterized by its input x and its output $y = f(x)$. A test $t(x,\ y,\ y_{exp})$ compares the output of the UuT with an expected output y_{exp}, resulting in

$$t(x,\ y,\ y_{exp}) = \begin{cases} success\ for\ y = y_{exp} \\ failure\ \ for\ y \neq y_{exp} \end{cases}$$

The real system f can be unknown or too complicated, so we use another system g to describe the properties of the UuT that are subject to the test:

$$y = g(x)$$

We expect g to behave similar to f.

We generate test cases for the real system f based on our known model g.

For test case generation, there are two approaches: Either selecting input vectors x_{gen} based on system knowledge or inverting the model g to compute input x_{gen} resulting in a desired output y_{exp}. The second approach can be implemented by analytically inverting the model (if possible) or using an optimizer:

$$x_{gen} = g^{-1}(y_{exp})$$

To execute our test t we use $x = x_{gen}$ as input for the system f.

If we cannot stimulate our system f directly with the input x (e.g. because our Unit under Test cannot be initialized with a certain state), an environment e has to be used. This environment transforms input x' to output x, which is then used as input for our system under test f:

$$x = e(x') \; or \; y = f(e(x'))$$

In case e is too complicated, we create a model h of this environment:

$$x = h(x') \; or \; y = g(h(x'))$$

Inverting our model h, we obtain the input:

$$\begin{aligned} x'_{gen} &= h^{-1}(x_{gen}) \\ &= h^{-1}(g^{-1}(y_{exp})) \end{aligned}$$

If the environment e (or our model of this environment h) is dynamic, i.e. it is time dependent, another aspect has to be considered. Using e with input x' (at time $t = 0$), we get the time dependent output:

$$x(t) = e(x', t)$$

The output $x(t)$ is used as input for our system under test f (and g):

$$y(t) = f(x(t)) = f\left(e\left(x', t\right)\right)$$

To test our system f, we generate the input x_{gen} by inverting g as shown above. We are interested in the input at given time t_{fix}. Therefore, we invert h to get $x'_{gen}(t_0) = x'_{gen}$ for input values x_{gen} and t_{fix}:

$$x'_{gen} = h^{-1}\left(x_{gen}, t_{fix}\right)$$

We can now test our system f embedded in e with the generated input x_{gen} and time t_{fix}:

$$\begin{aligned} y &= f\left(x_{gen}\left(t_{fix}\right)\right) \\ &= f\left(e\left(x'_{gen}, t_{fix}\right)\right) \end{aligned}$$

Due to the uncertainty emerging from the divergence between the models and reality, reaching the target states x_{gen} might happen not exactly at t_{fix} but within a range around t_{fix}. Therefore, we have to monitor if $y(t)$ shows the expected system response y_{exp} at least within a range around t_{fix}.

Our approach requires the inversion of g and h. Depending on the complexity of these models such inversion can be challenging and there are several general strategies of how to address this. The following chapter presents a possible approach to implement the inversion.

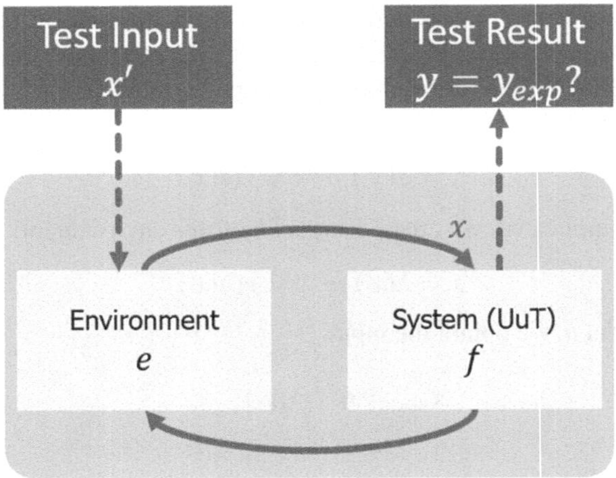

Fig. 2. Closed-loop test setup

The assertion of the test result $t(x, y, y_{exp})$ must not only be based on the comparison between y and y_{exp} but has to be checked additionally whether the target states x_{gen} are actually reached. A simple example illustrates this necessity: Imagine a test during which a vehicle drives at a velocity of $10\,{}^{\mathrm{m}}/_{\mathrm{s}}$ onto a target vehicle located 10 m ahead, and the expected system behavior y_{exp} is that the emergency brake is triggered. If the test is finished and no emergency brake has been triggered, we cannot state if the test failed or not. Due to a flawed environment model h the target vehicle might have been driving off so the target states x_{gen} could not be reached. Consequently, no emergency brake was triggered. On the other hand, if a crash occurred (i.e. the target states x_{gen} were reached), the missing emergency brake trigger shows that the test failed.

Figure 2 visualizes the relationship between the described system, the plant model and the test input.

3 System Inversion Using Equivalence Partitioning

As described in the previous chapter, our system is usually composed of various modules describing a certain part or function of an overall system. Inputs and outputs of these modules are well defined. An output of one module can be connected to an input of another module.

A test case of an isolated module would simply set the modules input and examine its output. As of the previously discussed necessity of realizing a closed-loop test setup utilizing a plant model, the test input x' parametrizes the overall setup.

To derive the input parameters x', the modules have to be inverted one after another. In our approach this results in a structured, machine-readable scenario description that can be automatically processed by the virtual vehicle created by the ETAS tooling.

The following workflow describes our approach by selecting input vectors x_{gen} based on knowledge of the system f:

1. Analyze the system f which has to be tested. The result of this analysis is the description of all relevant input dimensions d into the model g.
2. All identified input dimensions have to be subdivided into equivalence classes c_i^d. An equivalence class represents a range on an input dimension, in which every concrete selection of an input value causes the same behavior of system f (see [5]).
3. The resulting target states x_{gen} result by testing the combinatory of concrete values for each equivalence class within every dimension. If a single concrete value for each equivalence class is selected, the amount of resulting test cases is the result of multiplying the cardinality of the equivalence classes of all dimensions:

$$n_{Test\ Cases} = \prod_d \left| c^d \right|$$

4. It can be useful to use a selection of concrete values for each equivalence class. This results in a higher test coverage and enables utilization of best-practice approaches in software tests such as boundary-value analysis on equivalence classes [5]. As the simplified model g is only used for test case deduction, it is appropriate to apply value distribution on equivalence classes as the more complex system f is actually tested. This results in having a number of inputs x_{gen} to be set during the various test cases.

The result of this workflow is a set of scenario values x'_{gen} which describe the inputs of the environment model to reach the resulting target states x_{gen}. This set of scenarios parameterizes the environment model to reach the target states x_{gen} at a certain point in time. Then the reaction y of the system f can be observed. The test t is then examined by comparing this reaction with the expected reaction y_{exp}.

4 Use Case Realization

In this chapter we give an example on how our approach can be applied.

Our system under test f is an emergency brake assist (EBA). This function triggers an emergency brake based on the time to collision *ttc* and the relative position in y-direction d_y to a preceding vehicle. *ttc* is defined as the quotient of the relative distance in x-direction and the relative velocity v to the preceding vehicle:

$$ttc = \frac{d_x}{v}$$

The response of the EBA is either a brake request, a warning signal or no action:

$$f(ttc, d_y) = \begin{cases} \textit{no action} \\ \textit{warning signal} \\ \textit{brake request} \end{cases}$$

To efficiently describe the functionality of the EBA, we introduce the equivalence classes as follows:

$$c_0^{ttc} := \{ttc \in \mathbb{R}^+ \mid 0 \leq ttc \leq 1\}$$

$$c_1^{ttc} := \{ttc \in \mathbb{R}^+ \mid 1 < ttc \leq 3\}$$

$$c_2^{ttc} := \{ttc \in \mathbb{R}^+ \mid ttc > 3\}$$

$$c_0^{d_y} := \{d_y \in \mathbb{R} \mid 0 \leq |d_y| \leq 2\}$$

$$c_1^{d_y} := \{d_y \in \mathbb{R} \mid 2 < |d_y| \leq 2.2\}$$

$$c_2^{d_y} := \{d_y \in \mathbb{R} \mid |d_y| > 2.2\}$$

Based on these equivalence class definitions, the functionality of the EBA can be described as shown in Fig. 3 or on the left-hand side of Fig. 4. For our test case generation, we use a simple scenario with two vehicles on a straight road. Both vehicles are two meters wide, which leads to the choice of the equivalence classes of d_y. The vehicles

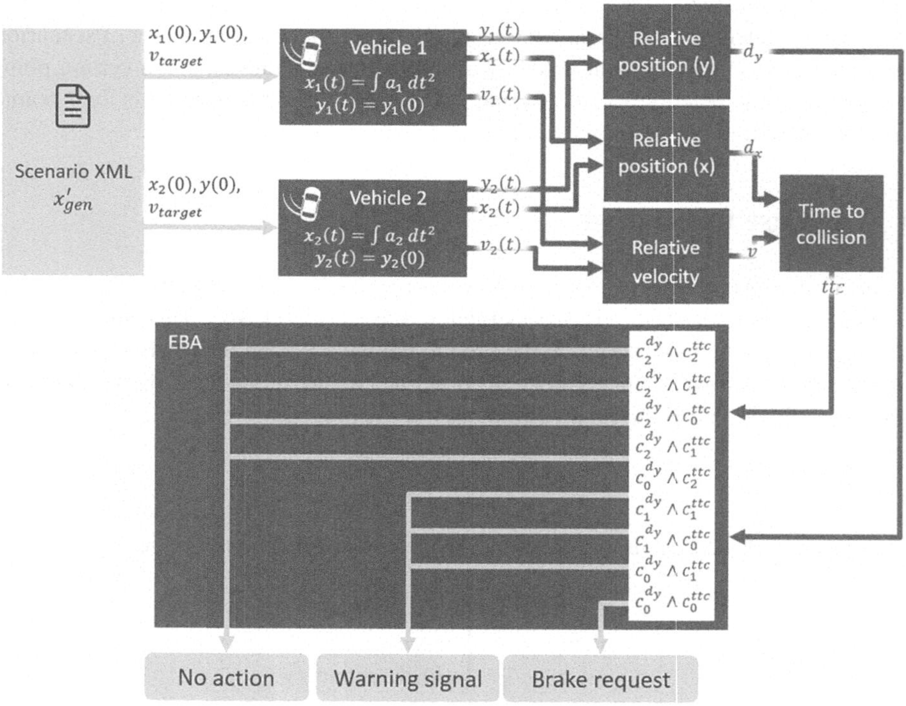

Fig. 3. Model to generate test cases for EBA

start accelerating without initial velocity. We use a scenario XML file, with a structure based on [2], to describe our driving scenario. This file can be parsed in a simulation.

To generate machine-readable test cases, we create a plant model which provides the input to the EBA and can be inverted to generate a scenario file based on the desired inputs of the EBA. We use two simple vehicle models with constant acceleration and without motion in *y*-direction. Furthermore, we create simple modules to compute *ttc* and d_y as input of the EBA model. The whole model is visualized in Fig. 3.

To generate test cases for each equivalence class, we have to invert our model for given values of *ttc* and d_y in order to obtain a scenario XML. We have the following underdetermined modules when inverting the whole system: "Time to collision", "Relative velocity", "Relative position (x)" and "Relative position (y)". In each of these modules, two input values lead to one output value. We always set one underdetermined input to a fixed value and calculate the other value. For all test cases, we choose:

- $d_x = 5$ m in "Time to collision"
- $v_2 = 3\,^{\mathrm{m}}/_{\mathrm{s}}$ in "Relative velocity"
- $x_2 = 300\,\mathrm{m}$ in "Relative position (x)"
- $y_2 = 0\,\mathrm{m}$ in "Relative position(y)"

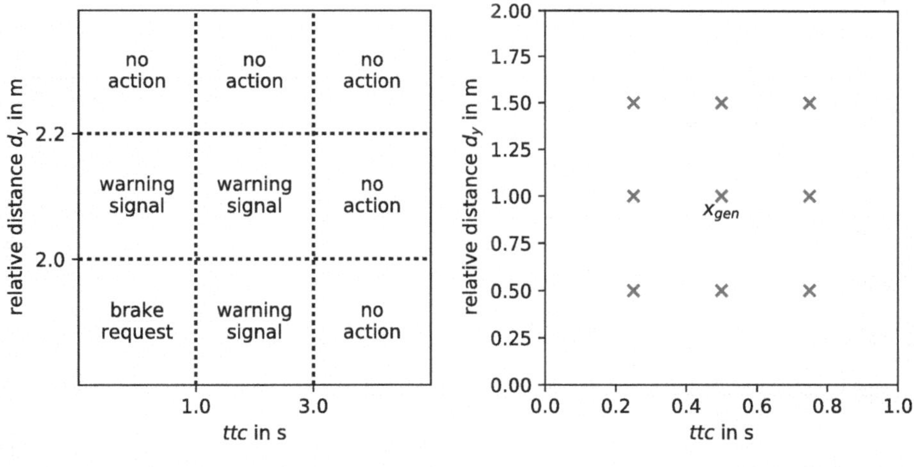

Fig. 4. Left: equivalence classes of ttc and d_y with resulting response of EBA, right: chosen discretization for generating test cases for response "brake request". x_gen is the selected test case for the simulation (see Fig. 7).

We now generate test cases for one combination of equivalence classes:

$$c_0^{d_y} \wedge c_0^{ttc}$$

All generated test cases are expected to lead to a brake request of the EBA:

$$y_{exp} = brake\ request$$

Since we are in general not completely certain that our choice of equivalence classes is correct (as stated earlier), we want to generate more than one test case for the above shown example. As representatives for the equivalence classes we choose an equidistant parameter distribution. Selecting $m = 3$ representatives per equivalence class leads to 9 target states x'_{gen} or respectively 9 generated test cases:

$$n_{testcases} = \prod_{\substack{d = \{d_y;\, ttc\} \\ i = 0}} \left(m \cdot c_i^d\right) = (3 \cdot 1) \cdot (3 \cdot 1) = 9$$

The chosen values in these two dimensions are shown in Fig. 4 on the right side.

The resulting values of these 9 test cases are shown in Fig. 5 for different levels while inverting the model. We can see that the chosen underdetermined values d_x, v_2, x_2, y_2 are equal for all generated test cases.

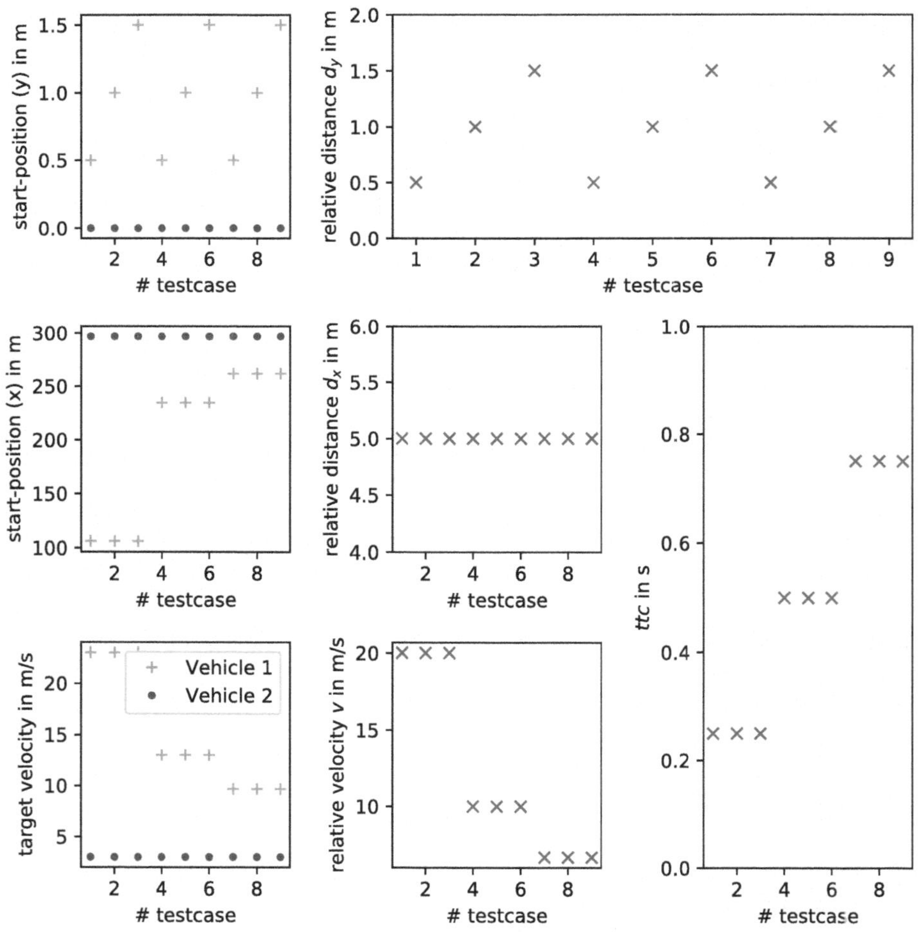

Fig. 5. Resulting values of the 9 generated test cases for different levels in the model.

5 Results

Finally, we compare the results of our model and a simulation for one of the 9 generated test cases above. We choose test case 5:

$$d_y = 1.0\,\text{m};\ ttc = 0.5\,\text{s}$$

We use a simulation framework developed by ETAS which can read a scenario XML and execute the simulation based on this file. For both vehicles, we use simple vehicle models having a constant acceleration. An overview of the simulation framework including the visualization of one of the generated EBA test cases is shown in Fig. 6.

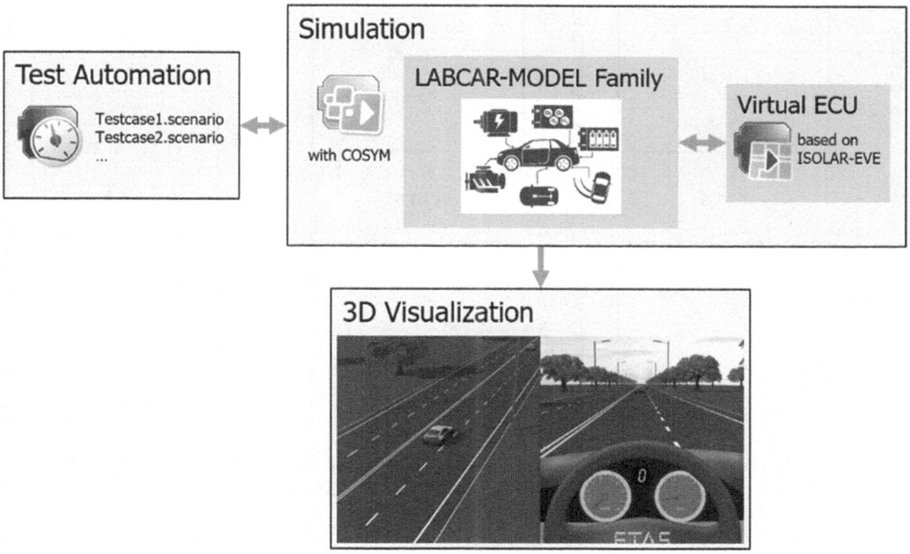

Fig. 6. Overview of the simulation framework.

The comparison of expected values of the model and values of the simulation for different levels of the model is shown in Fig. 7.

We can see, that the expected trajectory $(x(t), v(t))$ generated using an abstraction h matches the simulation e quite well. Furthermore, we observe, that the desired value of $ttc = 0.5\,\text{s}$ is never reached, because the EBA sends a brake request once $ttc < 1\,\text{s}$ and the vehicle breaks. Consequently ttc increases. This effect can also be seen in the relative velocity (decreases earlier as expected) and in the velocity of vehicle 1 (the target value is not reached because brake request occurs earlier). As visualized, the results are still within the accepted equivalence class $c_0^{d_y} \wedge c_0^{ttc}$. Therefore, the overall test executed as expected.

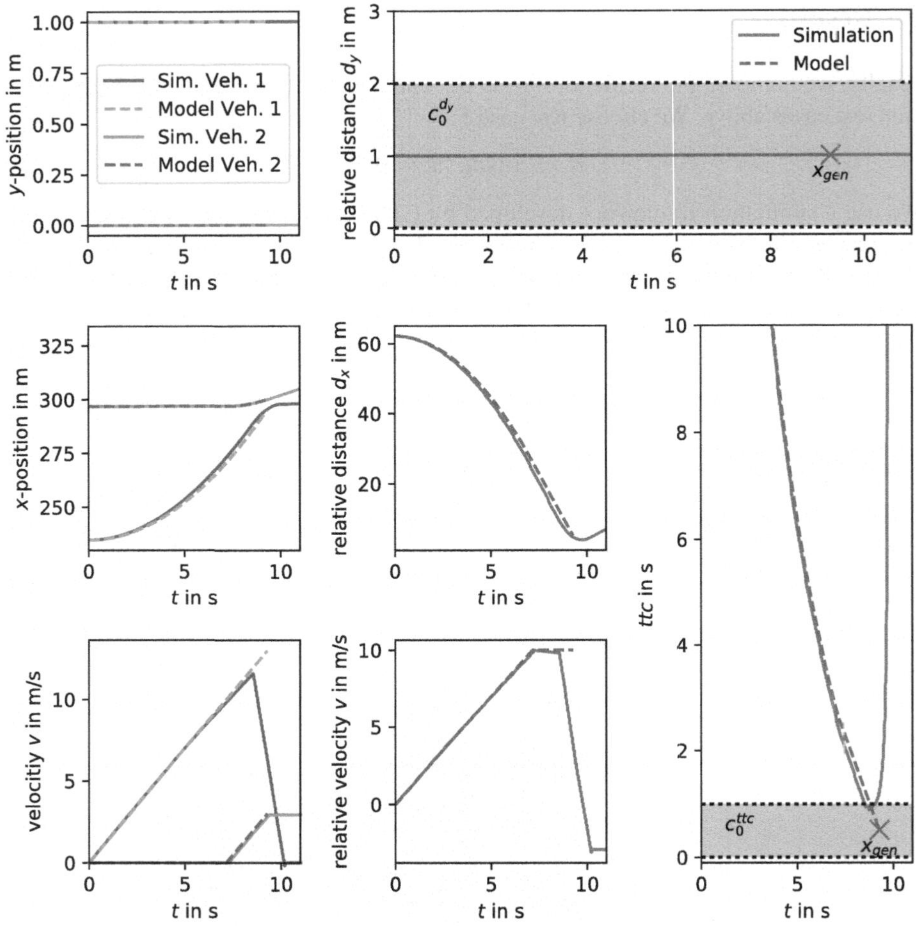

Fig. 7. Comparison of simulation and model expectations for different levels of the model. x_gen are the desired input values for the UuT (see Fig. 5).

6 Conclusion

In this work, we presented a framework to systematically generate machine-readable test cases. The generation is based on an abstract model of the system, which can be inverted. We achieve a complete test coverage by dividing all input dimensions of the system under test in equivalence classes. The generated test cases fully cover these equivalence classes. We showed our workflow of generating test cases for an EBA and compared the expected values with values obtained in simulation.

References

1. Donges, E.: Aspekte der Aktiven Sicherheit bei der Führung von Personenkraftwagen. Automobil-Industrie **27**, 183–190 (1982)
2. VIRES Simulationstechnologie GmbH: OpenSCENARIO. Bringing content to the road (2015). http://www.openscenario.org/, updated on 26/07/2017, checked on 23/01/2018
3. Feilhauer, M.; Häring, J.: Awendungsorientierte Absicherungsstrategie für Fahrerassistenzsysteme. In: ATZextra Automatisiertes Fahren. Springer Fachmedien Wiesbaden GmbH, Wiesbaden (2017)
4. Pütz, A., Zlocki, A., Eckstein, L.: Walting Absicherung hochautomatisierter Fahrfunktionen mithilfe einer Datenbank relevanter Szenarien. 11. Workshop Fahrerassistenzsysteme und automatisiertes Fahren. Uni-DAS. Walting im Altmühltal, 31 March 2017 (2017)
5. Liggesmeyer, P.: Software-Qualität. Testen, Analysieren und Verifizieren von Software, 2. Aufl. Spektrum, Akademischer Verlag, Heidelberg (2002)

Connected Development in Driver Assistance – Paving the Way to Automated Driving Through Remote Validation and Big Data Analysis

Tobias Radke[1(✉)] and Alberto Fernandez[2]

[1] Robert Bosch GmbH, Daimlerstr. 6, 71229 Leonberg, Germany
tobias.radke@de.bosch.com
[2] ETAS GmbH, Borsigstraße 14, 70469 Stuttgart, Germany

Abstract. The increasing complexity of driver assistance systems as well as the growing volume of data pose particular challenges to the development and validation of new systems. Connected Development is a new cloud-based approach which simplifies the validation process and shortens development iterations. Connected Development is an ongoing cycle, consisting of collecting data from the vehicle fleet and transferring it into a cloud storage, data management and analysis, writing new optimized software on request and software updates via remote flashing.

Even before the project starts, the engineers define which functions and driving situations should be monitored by Remote Validation. An on-board unit is required for transferring data to the cloud backend. While driving, surround sensors constantly capture the environment and process a multitude of data. If a defined situation occurs, such as Automatic Emergency Braking, only data from surround sensors categorized as relevant is captured by the on-board unit. The data is then uploaded to the cloud backend via a secure wireless connection practically in real time. At the same time and in the same manner, vehicle networking helps save data from locations around the world with different climatic conditions and from a range of vehicle configurations. The data center collects, organizes, checks and saves the recorded data from all networked vehicles and makes them available to the engineers. In this way, the engineers can effectively analyze the data from relevant driving scenarios. New software versions with optimized function performance and the corresponding validation jobs can instantly be developed based on the latest findings. After successful laboratory tests, the new software can be distributed securely and efficiently in the development vehicles via remote flashing.

T. Radke—Senior Project Manager Connectivity for Driver Assistance.
A. Fernandez—Director Product Management Measurement, Calibration and Diagnostics Solutions.

T. Bertram (Hrsg.): *Fahrerassistenzsysteme 2018*, Proceedings, S. 126–131, 2019.
https://doi.org/10.1007/978-3-658-23751-6_13

Connected Development enables shorter learning cycles, increased efficiency and guaranteed quality in the development of new driver assistance systems. The field data exploration continues this approach even after serial production begins. In this case, it serves as a key enabling technology for the validation of future highly automated driving functions as well as for new data-based services such as Predictive Diagnosis and Maintenance.

1 Motivation

Development and validation of new driver assistance and automated driving systems are currently facing three major challenges (Fig. 1):

1. Strong increase in validation effort
2. Strong increase in variants
3. Strong increase in complexity

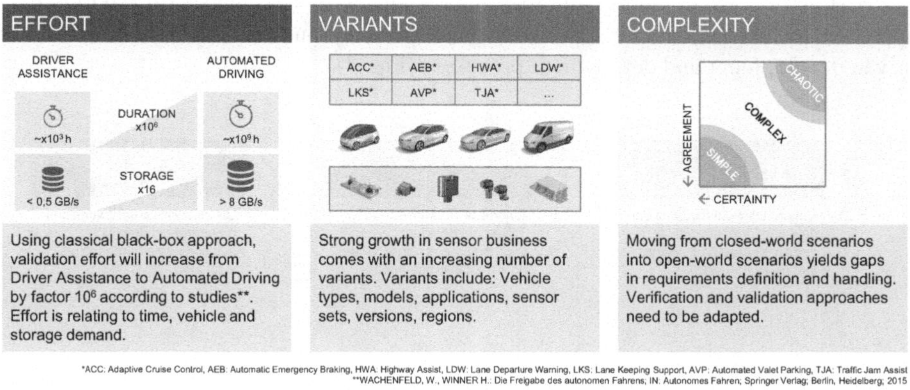

Fig. 1. The validation challenge

These challenges need to be addressed to guarantee time-to-market and quality for new systems and functions. Particularly challenging is the shift from Advanced Driver Assistance Systems (SAE Level 1–2), in which the driver monitors the driving environment, to Automated Driving Systems (SAE Level 3–5), in which the system (at least temporarily) takes over the monitoring. This shift results in open-world problems, i.e. previously unknown scenarios have to be handled properly by the system. Advancing from "Known known" and "Known Unknown" scenarios to successive coverage of "Unknown unknown" scenarios requires adaptations in the verification and validation approaches (Fig. 2).

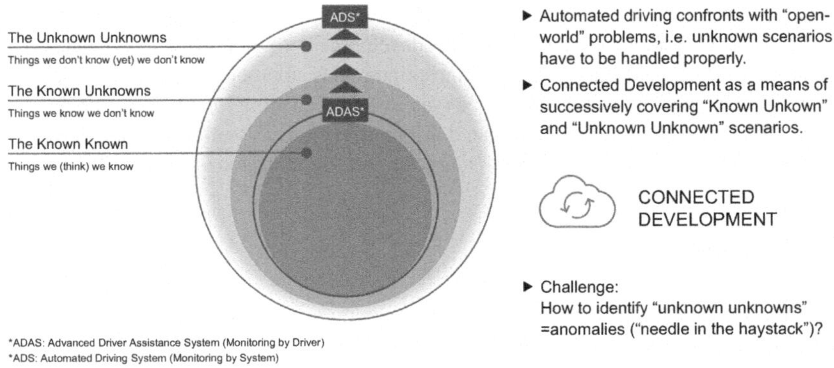

Fig. 2. The validation challenge – "Unknown Unknowns"

2 Connected Development Approach

BOSCH and ETAS are addressing these challenges with a "Connected Development" approach to raise validation effectivity and efficiency and to support iterative data-driven development and deployment processes (Fig. 3).

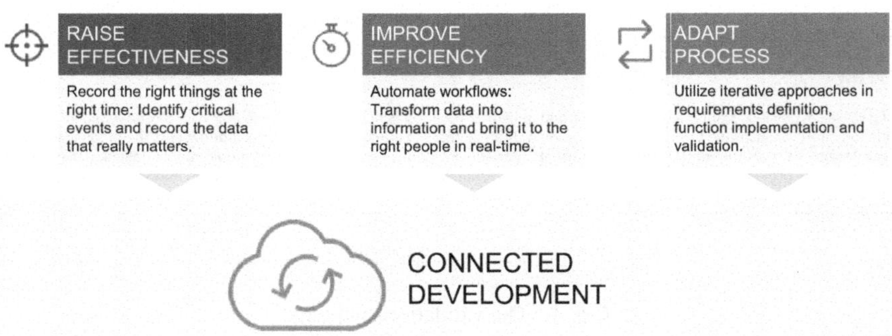

Fig. 3. The validation challenge – Targets

Connected Development is an ongoing cycle, consisting of collecting data from the vehicle fleet and transferring it into a cloud storage, data management and analysis, writing new optimized software on request and software updates via remote flashing.

Even before the project starts, the engineers define which functions and driving situations should be monitored by Remote Validation. An on-board unit is required for transferring data to the cloud backend. While driving, surround sensors constantly capture the environment and process a multitude of data regarding lane markings, pedestrians and objects. If a defined situation occurs, such as Automatic Emergency Braking, only data from surround sensors categorized as relevant is captured by the on-board unit. The data is then uploaded to the cloud backend via a secure wireless connection practically in real time (Fig. 4). At the same time and in the same manner, vehicle networking helps save data from locations around the world with different climatic conditions and from a range of vehicle configurations. The data center collects, organizes, checks and saves the recorded data from all networked vehicles and makes them available to the engineers. In this way, the engineers can effectively analyze the data from relevant driving scenarios. New software versions with optimized function performance and the corresponding validation jobs can instantly be developed based on the latest findings. After successful laboratory tests, the new software can be distributed securely and efficiently in the development vehicles via remote flashing.

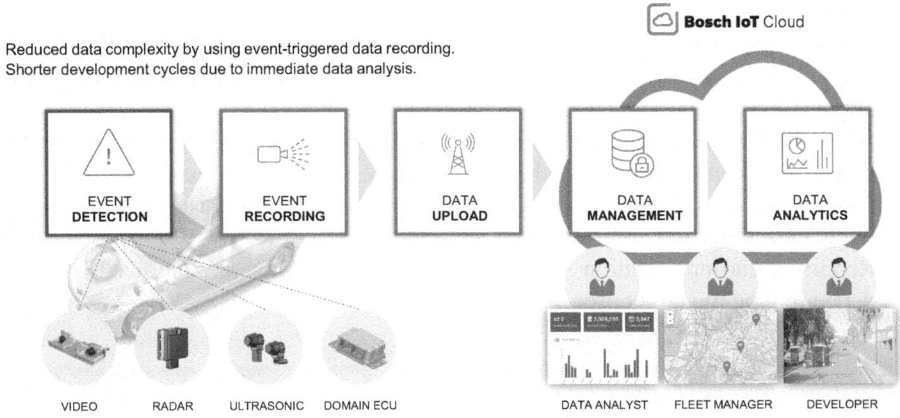

Fig. 4. Connected validation methodology

Automatic Emergency Brake (AEB) validation gives a good example of possible Connected Validation features and benefits. In this case, a Point of Interest map can be aggregated automatically in the cloud backend from the data recorded and utilized to plan validation runs in an optimally effective manner. Efficiency on the other hand can be increased by concentrating on the immediate wireless transmission of critical or sub-critical events with only its most relevant data (Fig. 5).

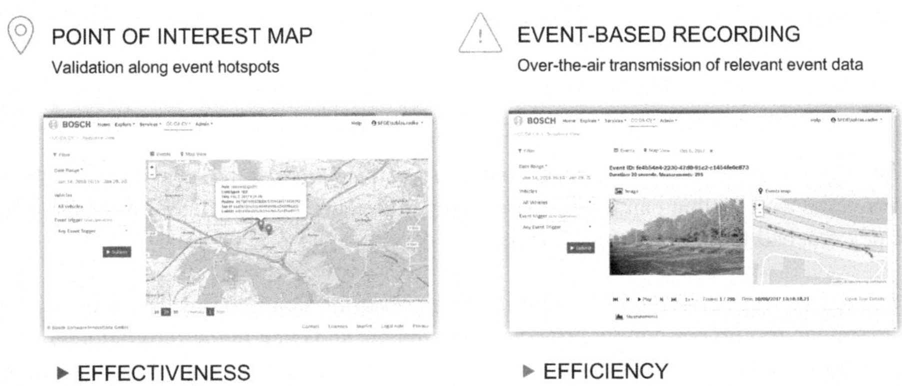

Fig. 5. Example – Automatic Emergency Brake (AEB)

3 Infrastructure Components

To realize Connected Validation several infrastructure components are necessary (Fig. 6):

1. An end-user frontend software accessible via web-browser technology to ensure a global access to the system.
2. A backend infrastructure including the software that enables the targeted use cases (e.g., remote flashing, remote measurement). Data will be also received, stored and managed in the backend infrastructure. Such an infrastructure ensures scalability of the system.
3. Secured data channels. Both over the air, to quickly interact between backend and vehicle fleet, and via cable to transmit high data volumes if required (e.g., for autonomous driving use cases).
4. An on-board unit connected to the relevant ECUs and sensors. Depending on the data acquisition requirements, this system may contain several hardware units.
5. Data-analytics software to index, manage and process data according to predefined algorithms and criteria.
6. Reporting and presentation software to automatically generate the needed views depending on the target user.

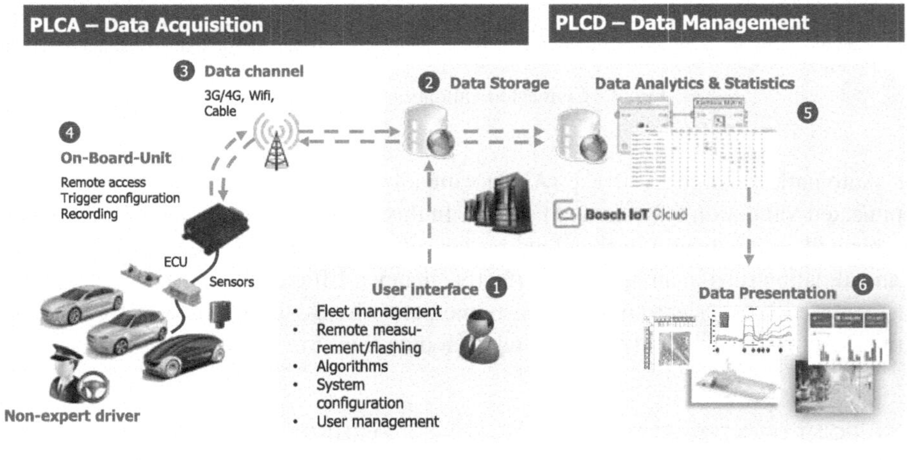

Fig. 6. Necessary infrastructure components

Connected Validation is being rolled out today into productive use. Nonetheless there are several aspects that can be further developed to increase the effectiveness and efficiency reached with this approach (Fig. 7):

7. Provide a scalable hardware on-board unit portfolio that maximizes the sharing of vehicle prototypes among projects, departments and even in the future domains.
8. Minimize the data traffic (especially over the air) with more complex preprocessing of the data in the on-board hardware system.

9. Ensure that data is globally available to all teams and functions within companies.
10. Create higher value added out of the available data with new artificial intelligence technologies.

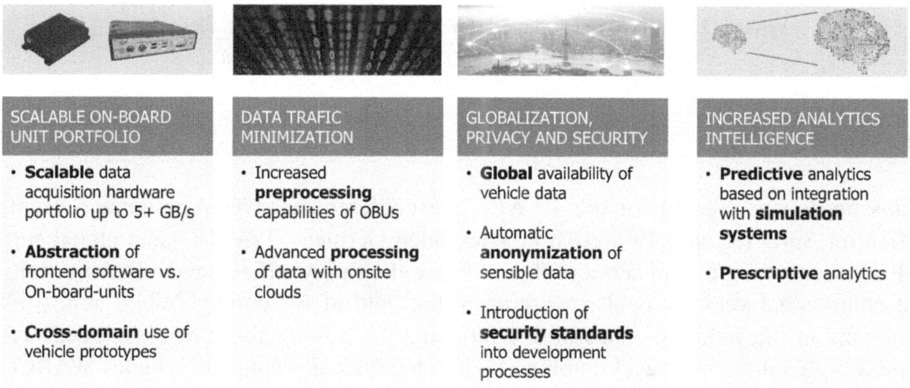

Fig. 7. Infrastructure – Main working areas for next generations

4 Outlook: Edge Computing Approach Across Component Portfolio

On an abstract level, Connected Development can be seen as an Edge Computing approach applied to the automotive validation domain: With driver assistance sensors creating larger and larger amounts of data with every new generation, it seems essential to pre-process the majority of data "on the edge", i.e. in the vehicle, instead of streaming everything to the cloud backend in raw format. To accomplish this, intelligence needs to be shifted from the cloud backend to the vehicle. BOSCH follows this approach by successively implementing smart data pre-processing on the relevant ECUs across the component portfolio (Fig. 8).

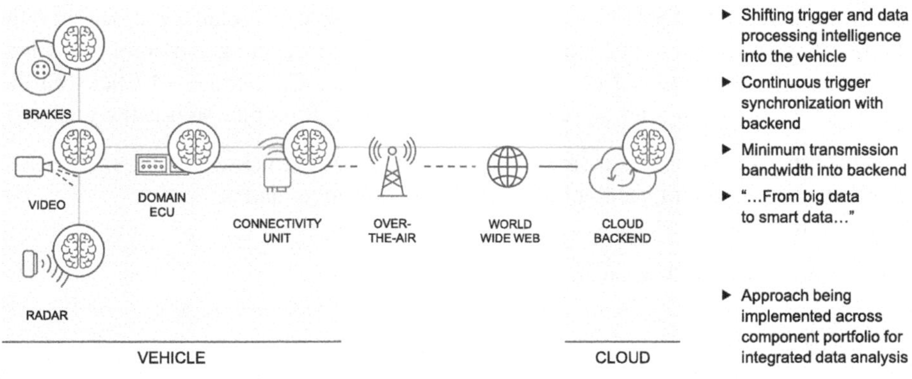

Fig. 8. Outlook – Edge computing approach across component portfolio

Truck Platooning – A Pragmatical Approach

Stephan Kallenbach(✉)

WABCO GmbH, Am Lindener Hafen 21, 30453 Hannover, Germany
stephan.kallenbach@wabco-auto.com

1 Introduction

This paper was created for the 4th ATZ live conference Driver Assistance Systems 2018 on April 18th and 19th, 2018 in Wiesbaden Germany. WABCO as a global supplier of technologies and services that improve the safety, efficiency and connectivity of commercial vehicles is also working in the field of Platooning. While almost all attempts in the industry concerning Platooning are aiming for very close headways and a high automation level combined with electronically coupled vehicles WABCO is additionally investigating a slightly different Platooning approach, which might be easier to realize and might have higher acceptance amongst people.

1.1 Scope

The goal of this work is to give a view into the WABCO's perspective on Platooning. It will show the results of an engineering based pragmatic investigation of Platooning and its major influencing factors and some conclusions are derived from further steps.

1.2 Truck Traffic Situation on Today's Roads

The truck density on European roads is increasing continuously based on economical growth and the fact, that more and more goods are transported by trucks. Road infrastructure is not growing the same which results in rising traffic congestion. Due to this also rear end collisions, meaning trucks crashing into traffic jams, are happening more often. These crashes often result in at least severe injuries or even fatalities. In addition this high number of trucks and traffic jams are causing an increased amount of emissions which should be reduced in times of global warming and climate change.

The simplest solution of course would be to reduce the number of trucks on the roads, but due to transportation requirements, this is not possible. So it is the goal to

- reduce the emissions, i.e. increase the fuel efficiency of trucks,
- reduce the number of accidents, i.e. increase road safety and also,
- reduce the space trucks are occupying on motorways.

Truck Platooning could be one option.

T. Bertram (Hrsg.): *Fahrerassistenzsysteme 2018,* Proceedings, S. 132–157, 2019.
https://doi.org/10.1007/978-3-658-23751-6_14

1.3 Definition Platooning

Basically a Truck Platoon is a convoy that is two trucks or more driving behind each other in the same direction. The following distance between the Platoon members is electronically controlled. So in general Platooning can be compared with an Adaptive Cruise Control (ACC). Most Platooning definitions include an electronic coupling between the Platoon members based on a Wi-Fi protocol (IEEE 802.11p). Via this wireless link the Platoon members can exchange data which means that the vehicle controller is not based on a forward looking sensor only, like in today's ACC systems, but also has direct access to the preceding truck. This Wi-Fi link reduces the system reaction time compared to an ACC which is based on RADAR only. If the driver depresses the brake pedal for instance, in a Platooning system this signal can be transmitted over the air to the following truck, so that it immediately can react on this brake action. In a RADAR based system the lead vehicle has to decelerate first and this deceleration has to be detected by the RADAR before an ACC in the following truck can react on it.

Due to the fact that the system reaction time can be reduced by this vehicle-to-vehicle (V2V) connection the following distance between the Platoon members can be reduced. Concerning the headway between Platooning members many different values exist. The range lasts from 0.3 s (7 m @ 89 kph) over 0.5 s (12 m @ 89 kph) to 0.8 s (20 m @ 89 kph)[1]. This headway reduction shall improve the fuel efficiency due to drag reduction for the following truck. This is because the trailing vehicle is in the wake of the preceding one, which is also called *drafting*. On very close headways also the preceding truck has a positive effect because the amount of reduced pressure behind the truck (suction area) is minimized (refer to Table 11, Appendix A.6 for some examples of time gaps dependent on speed).

On the other hand the reduced headway also saves road space. A truck platoon occupies less space due to the reduced following distance (see Fig. 1).

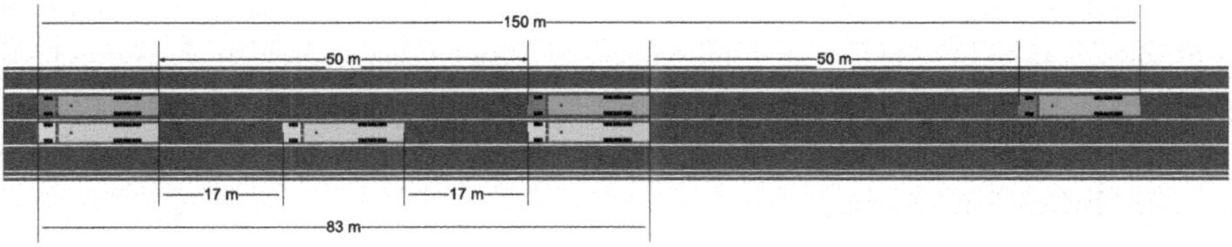

Fig. 1. Example for a three truck platoon, just for illustration. Three trucks are occupying the same space as two trucks with 50 m legally required safety distance. (17 m equals a time gap of approx. 0.7 s at 89 kph.)

In a Platoon at least the longitudinal dynamics must be controlled automatically (SAE level 1[2]) but there are also concepts were also the lateral movement of the vehicle

---

[1]Information: Legal requirement for following distance between trucks is 50 m or 2 s (50 m @ 89 kph) in many European countries. Even if there is no mandated headway, the 2 s rule is applied in court.

[2]Refer to SAE J3016 for level of driving automation definition (simplified version see Fig. 13).

is controlled (SAE level 2). In both scenarios the driver shall be the fall back solution in case the system fails which is at least questionable, if a human being is able to take this role. Truckers today already suffer under the monotonous driving task. Platooning will not improve this situation; the driver's visibility for instance is blocked by the preceding truck.

Based on the V2V communication the following truck copies the throttle and brake requests of the preceding truck. This automation increases the road safety simply because less rear end collisions will occur.

2 Fuel Savings

One reason for Platooning is to save fuel based on drag reduction due to the close following distance and therefore driving in the wake of the lead vehicle. There have been many investigations on this topic already. In general the results always state a reasonable fuel saving rate dependent on close following distances. The closer the distance the higher the fuel saving rate. Another similarity these results show is, that either only close following distances were investigated, or that these were the only ones published. Moreover, the fuel saving rate was always related to a single driving truck. As stated above, the situation of a single driving truck on a European highway is not the standard. WABCO therefore was interested in the fuel saving rate of larger headways, especially those which are driven by truckers today and the mandated following distance of 50 m or 2 s.

Since these values were not available WABCO conducted own fuel saving tests. These will be presented within this chapter and compared to fuel saving tests already released by other parties.

Most of the already published Platooning tests were done with two trucks, i.e. one leader and one follower. If the results were based on a three truck platoon, i.e. two trailing trucks (NRC [1] and Japan ITS) the fuel savings for the follower were taken from the 3rd vehicle which is not influenced by another one behind it. It is assumed that this can be compared best with a two-truck platoon.

2.1 Published Fuel Savings

For fuel savings dependent on Platooning mainly the portion of the air resistance is of importance. This depends on the vehicle speed v, the air drag coefficient c_W and the front area of the truck A (Eq. (1)).

$$F_A = \frac{1}{2} \cdot c_W \cdot A \cdot \rho \cdot v^2 \tag{1}$$

The power which is needed to overcome the air resistance is

$$P_A = F_A \cdot v = \frac{1}{2} \cdot c_W \cdot A \cdot \rho \cdot v^3 \tag{2}$$

The results will be split into North American (NA) style vehicles (long-hood trucks) and cab-over-engine trucks (COE or cabovers) which are used in Europe and Asia. Cabin shapes and length differ considerably (Fig. 2). In addition the maximum vehicle speed driven with cabovers is 90 kph (56 mph) whereas the speed in North America is much higher (range between 89 kph (55 mph) and 137 kph (85 mph) dependent on the state).

Fig. 2. Left: typical North American tractor (long-hood); Right: typical European tractor (cab-over-engine)

Except for the WABCO own results all numbers were taken from NRC [1], NACFE [2], Peterbilt/EXA [3] and NREL [4]. As already stated those results are only dealing with very close distances and do not consider larger headways.

2.1.1 North American Vehicles

Figure 3 shows published fuel savings for the lead (LV) and the following vehicle (FV) (detailed values can be found in Table 7, Appendix A.2). The lead vehicle does not benefit from a trailing truck until the following distance gets really close. To achieve savings of 2% you need to get closer than 0.8 s and for 4% closer than 0.4 s.

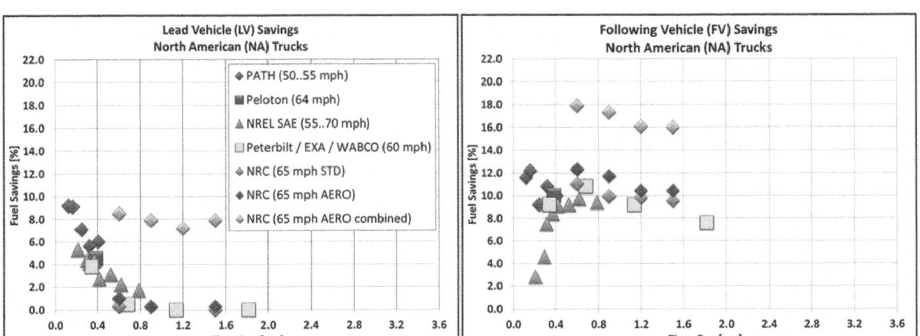

Fig. 3. Fuel savings North American vehicles; Left: Platooning fuel savings for leading vehicle (LV); Right: Platooning fuel savings for following vehicle (FV)

For the following vehicle the situation is completely different. Based on published results it seems that the trailing truck already benefits from larger headways. NRC tested up to 1.5 s headway and Peterbilt/EXA simulated up to 1.8 s. Fuel savings of 6 to 8% seem to be achievable at headways considerably greater than 1 s. NRC indicates fuel savings of 6.7% at 1.5 s headway [1] and the CFD simulation presents the prospect of 7.6% at 1.8 s [3]. Based on the CFD simulation it looks like the fuel savings – respectively the positive draft effect – are not increasing infinitely by reducing the time gap. The rising trend starting from 1.8 s stops below 0.7 s. It shall be noticed that these results are based on aerodynamic effects only and do not consider fuel consumptions due to other effects. Also the NREL [4] results does not show a continuous positive effect after undershooting 0.8 s. Moreover there is a remarkable drop after falling below 0.4 s. In this case it is caused by the engine fan which becomes active at such close headways as the following truck is not getting enough ram air to cool the engine. It can be concluded that there is an optimal headway for the trailing truck where it benefits from drafting as well as getting enough air for engine cooling.

Besides the standard configuration (*STD*), NRC also conducted tests with aerodynamic trailer devices like side-skirts or boat-tails (*AERO*). In case of *AERO* all trailers (control and platooning trailer) were equipped with skirts and tails. In case of *STD* no trailer hat any aerodynamic device installed. The difference between *STD* and *AERO* is not very high caused by the fact, that these are relative values based on a *STD* or *AERO* reference. That means that both results show the pure Platooning effect, which is comparable (approx. 10 to 12%). NRC also calculated the savings due to Platooning and aero devices (*AERO COMBINED*) which yield another 6% of savings. Based on these results an aero platoon saves approx. 16 to 18% fuel compared to a single driving truck trailer combination without any skirts and tails installed on the trailer. For the leading vehicle the aero benefit is approx. 8%.

2.1.2 Cab-Over-Engine Vehicles

The fuel saving results for the lead truck in case of cabovers is similar to NA vehicles. There is no fuel saving benefit for larger following distances. The effect even seems to start later than with NA trucks. At 0.8 s still no positive effect was observed at cabovers whereas long-hood trucks already achieve approx. 2% (compare Sect. 2.1.1).

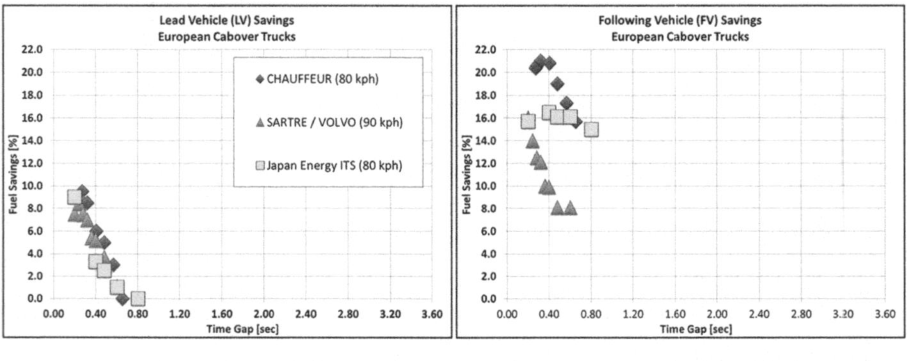

Fig. 4. Fuel savings cabover vehicles; Left: Platooning fuel savings for leading vehicle (LV); Right: Platooning fuel savings for following vehicle (FV)

Looking at the platoon follower and comparing it to the NA results the gradient is much steeper. It looks like a linear dependency between following distance and savings. CHAUFFEUR and SARTRE characteristics are similar whereas the Japan ITS values are almost constant. The offset between the result might be caused by different reference the percentage is based on. The tendencies however are more or less identical (Fig. 4).

2.2 WABCO Fuel Saving Results

As stated above WABCO has conducted own fuel saving tests for two-truck Platooning. With NA vehicles in the USA as well as with cabovers in Europe. In both cases the test specification was based on SAE J1321 [5].

Our main goal was to verify and compare own measurements with the published ones and additionally to gather data for larger following distances which were not tested yet by other parties.

2.2.1 North American Vehicles

Tests were conducted at TRC, Ohio in September 2017. Three identical Kenworth T680 6x4 tractors and three identical Hyundai dry van 53 ft trailers were used. All combinations were loaded to 29,400 kg (65,000 lbs). According to [5] the fuel savings were determined based on fuel balance. The vehicle test speed was 105 kph (65 mph). Two configurations were tested:

- Without aero improvement parts at all trailers, tested headways: 3.6, 1.4, 0.9 and 0.5 s.
- With aero improvement parts skirts and boat tail (WABCO OptiFlow™) installed on the trailer of the platoon follower, tested headways: 3.6 and 1.4 s.

In addition WABCO added an additional time gap of 3.6 s to the CFD simulation done by Peterbilt and EXA in the first place [3] and repeated the simulation.

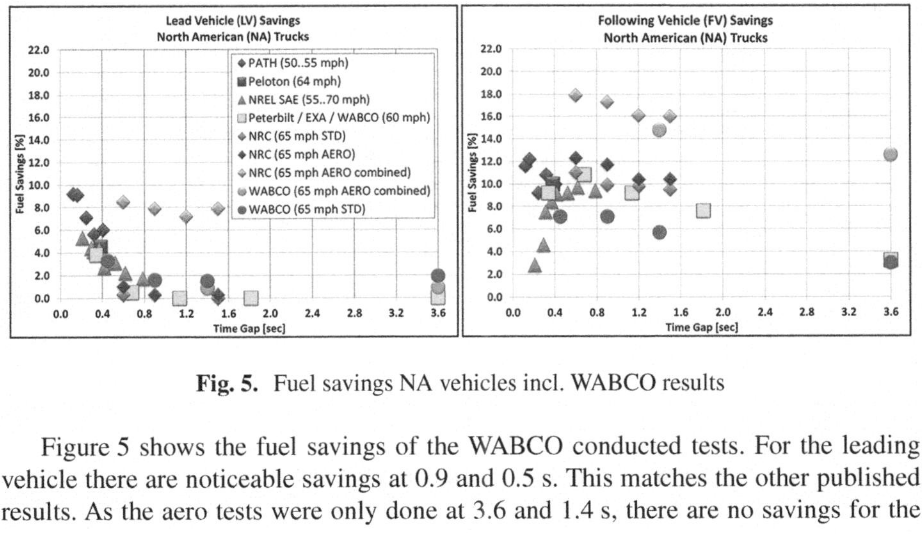

Fig. 5. Fuel savings NA vehicles incl. WABCO results

Figure 5 shows the fuel savings of the WABCO conducted tests. For the leading vehicle there are noticeable savings at 0.9 and 0.5 s. This matches the other published results. As the aero tests were only done at 3.6 and 1.4 s, there are no savings for the platoon leader. In general the following vehicle results are lower than the results of prior publications but still indicate savings of approx. 6% for a headway of 1.4 s. The 3.6 s distance matches the result of the additional CFD simulation at 3.6 s (3%). Also the tests conducted with the aero devices match the trend of the NRC tests. The result is that aero devices like WABCO OptiFlow™ are not interfering with the Platooning effect but have an additional positive effect on it. A first fuel efficiency step therefore could be to use side-skirts and boat-tails for all trailers and use Platooning in a second step.

2.2.2 Cab-Over-Engine Vehicles

Tests were conducted at the RDW testcentrum in Lelystad, Netherlands. The first test session was conducted in January 2017 and the second one in August 2017. Three identical DAF XF 4x2 tractors and three identical Krone dry box trailers were used. All combinations were not loaded so the gross vehicle combination weight was approx. 14 t. Test configuration was according to [5] except for the fuel savings which were not determined by gravimetric fuel weight but based on CAN data. The vehicle test speed was 85 kph (53 mph). Tested time gaps during both sessions were: 0.4, 0.8, 1.3, 1.7 and 2.1 s which equals 10, 20, 30, 40 and 50 m at 85 kph. In the summer session one additional configuration 3.0 s (70 m) was tested. The reason for not testing larger distances as it was done in the US tests, is that the mandated following distance in Europe is either 2 s or 50 m (refer to Table 10 in Appendix A.5).

The fuel savings for the platoon leader are between 3 and 4% and only measureable at 0.4 s. This result matches prior Platooning publications. The results for the platoon follower were highly interesting since there are no comparable results for headways greater than 0.8 s available in present publications. The tests showed that even at the current mandated distance of 50 m/2 s the savings for the follower are between 7 and 8%. Moreover the savings are relatively constant over the whole range. The WABCO results do not show a characteristic increase as the CHAUFFEUR and SARTE values do. Aside from the offset the WABCO characteristics are similar to the Japan ITS values. The offset might arise from a different reference the percentage is based on (Fig. 6).

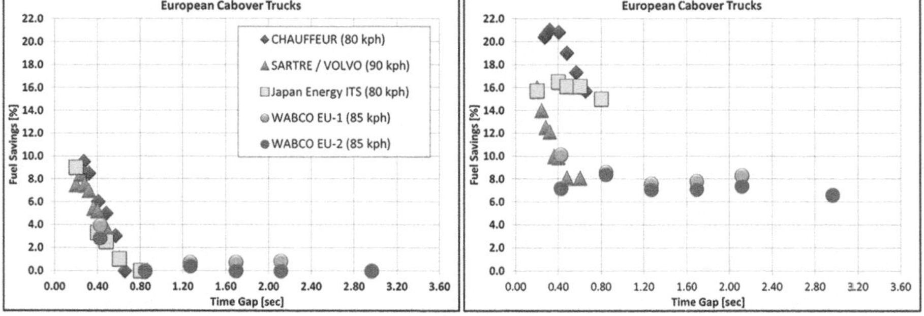

Fig. 6. Fuel savings cabovers incl. WABCO results

2.3 Conclusion of Fuel Savings

The lead truck only saves fuel at very short headways of the following vehicle. This is valid for NA as well as for cabover vehicles in the rest of the world. The threshold is approx. 0.8 s. To gain savings of more than 2% you need to get closer than 0.5 s. The WABCO tests attained identical results. It is at least questionable if fuel savings of 2% are justifying the effort which is needed to realize such short headways.

The main goal was to determine the fuel savings for the following truck at larger distances. For NA trucks the results show noticeable savings (6 to 9%) for time gaps clearly larger than 1 s. Similar results could be obtained for EU cabovers. Savings were 6 to 8% even for a time gap of 2.1 s (50 m at 85 kph).

Based on these results the conclusion is that, already today truck drivers are saving fuel due to drafting at the mandated distance of 2 s or 50 m. Getting closer and closer to the preceding truck does not seem to have a noticeable positive effect on the savings. Moreover, the shorter the distance the greater the problems concerning missing air flow to cool the engine. Also an aerodynamic optimization of a platoon is not yet investigated. Current trucks and trailers are not optimized for driving close behind each other.

2.4 Infinite Convoy of Trucks

Based on the fuel saving results a very interesting platoon is a convoy with not too close headways where everybody is follower. By this it is not necessary to build up platoons of two, three or four trucks etc. and have a dedicated leader who has nobody in front or needs to have a larger headway to the next truck. Since the inter vehicle distances are not too close there wouldn't be any problem at bridges, tunnels or entrance and exit ramps. Such convoys would deliver fuel savings now and could be pre stage to a later Platooning with closer headways.

3 Rear-End Collisions

One topic that often comes into play is the risk of rear end collisions and in case of Platooning the collision between platooning members in case of an emergency braking. V2V communication is always mentioned as the solution to overcome the long driver reaction time and also the system reaction time of camera and/or RADAR based distance control systems. The time delay of less than 0.1 s for pure radio communication might be very low compared to driver reaction times of 1.25 s (average Kölner Modell [6]) but it is still a delay which causes a lag in the deceleration of the vehicles and therefore uses up the inter vehicle distance. Also the duration of data processing in the lead and trailing vehicle, the encryption of data, the system reaction time of the brake system needs to be considered. The distance wasted due to the delayed start of deceleration in the trailing vehicle cannot be compensated unless the trailing vehicle decelerates stronger than the preceding one, but this is not possible under all circumstances.

3.1 Influencing Factors of Deceleration Capabilities

Driving in a platoon with short headways and avoiding rear end collisions requires both fast reaction times of the trailing vehicles and the capability of the followers to realize the needed deceleration. Almost simultaneous braking does not help if the deceleration of the leader cannot be matched by the follower. Besides the time delay or reaction time of the braking signal transmitted from the leader to the follower influencing factors are:

- Inhomogeneous vehicles in terms of brake system, means of communication and distance control system.
- Based on WABCO experience known deceleration performances of trucks are located in a range between 4.5 and 8.5 m/s^2 resulting in a deceleration differential of 4 m/s^2. The legal requirement in Europe (UNECE-R13) mandates a minimum deceleration of 5 m/s^2 for a single truck and 4.5 m/s^2 for a tractor semitrailer combination.
- Differences in brake performance are influenced by:
  - Tires, not just wear and tear, aging and tire pressure but also make and model of the tires. Based on the rubber compound a tire has different characteristics on dry and wet surfaces.
  - Road surfaces and conditions, such as μ-split and μ-jump conditions, wheel ruts, etc.
  - Brake system condition, such as wear, temperature, reaction and rise time, supply pressure level, maintenance, aging, simply differences dependent on make and model of the friction brake parts (discs or drums and linings).
  - Key indicators like road friction coefficient μ cannot be determined in the needed accuracy before start of braking but only during braking.
- Also the trailer needs to be taken into account. All above mentioned factors also need to be considered for the used trailer. Trailers are independent vehicles and might be exchanged between trucks and tractors on a regular basis. Moreover it shall be noticed that no truck OEM manufactures trailers and vice versa.

The next chapter will show, that above mentioned factors have a major impact on the probability of rear-end collisions than the transmission time lag.

3.2 Calculation of Stopping Distance

To ascertain whether a rear-end collision between platoon members will occur due to an emergency braking done by the lead vehicle the overall stopping distances of both vehicles have to be compared. The emergency braking of the lead vehicle is either caused by a driver initiated full panic braking or by an autonomous braking system like an AEBS. For the calculation of the complete brake distance the initial following distance, the time delays due to data transmission and processing, the brake rising times etc. need to be considered. If the calculated delta between the needed distances of both trucks is greater than 0 the collision was avoided.

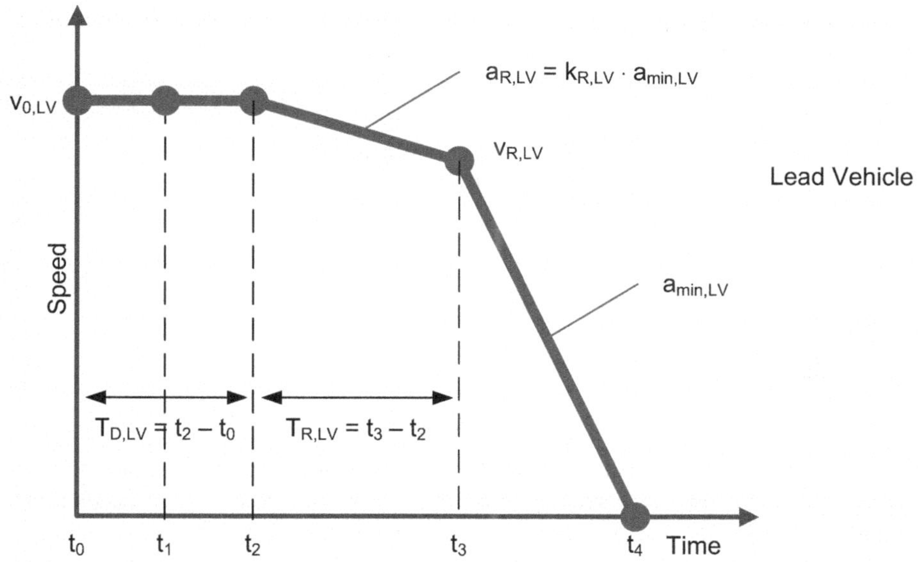

Fig. 7. Full braking characteristics of the lead vehicle (LV) (not true to scale)

Since it is a complex calculation depending on many factors the following simplifications have been done: Fig. 7 shows the characteristics of the vehicle speed over time during a full braking. At t_0 either the driver fully depresses the brake pedal or an autonomous system transmits a brake request to the brake system. At t_1 the brake system EBS receives the brake request and calculates the needed pressure per axle dependent on load etc. This axle pressure is then transmitted to the respective axle modulators and the trailer. The time duration between t_2 and t_0 is the so called dead time T_D whilst the vehicle is not yet braking.

$$T_D = t_2 - t_0 \tag{3}$$

At t_2 the brake rising time starts, i.e. the brake valves are opened and the air flows from the supply reservoir to the brake chambers and builds up the needed brake pressure.

$$T_R = t_3 - t_2 \tag{4}$$

For simplification the deceleration during brake rising time a_R is assumed to be constant. It is calculated based on the maximum deceleration capability of the truck a_{min} as follows:

$$a_R = k_R \cdot a_{min} \quad (0 < k_R < 1) \tag{5}$$

The overall stopping distance s is the sum of the distance s_1 driven during dead time T_D without any deceleration, distance s_2 driven during brake rising time where the vehicle decelerates with a_R and s_3 the full braking phase with maximum deceleration a_{min}.

$$s = s_1(T_D, v_0) + s_2(T_R, v_0, a_R) + s_3(v_R, a_{min}) \tag{6}$$

The single distance s_i (for $i = 1, 2, 3$) are calculated as follows. The derivation of the equations is renounced

$$s_1 = v_0 \cdot T_D \tag{7}$$

$$s_2 = v_0 \cdot T_R + a_R \cdot \frac{T_R^2}{2} \tag{8}$$

$$s_3 = -\frac{v_R^2}{2a_{min}} \tag{9}$$

in which the speed v_R is

$$v_R = v_0 + a_R \cdot T_R \tag{10}$$

The calculation of the overall stopping distance is identical for the lead and the following vehicle but time lags such as brake rising time or the maximum deceleration capability will be different. In addition the following vehicle will not brake simultaneously with the lead vehicle. The platoon follower will therefore experience a longer lasting dead time $T_{D,FV}$ (Fig. 8).

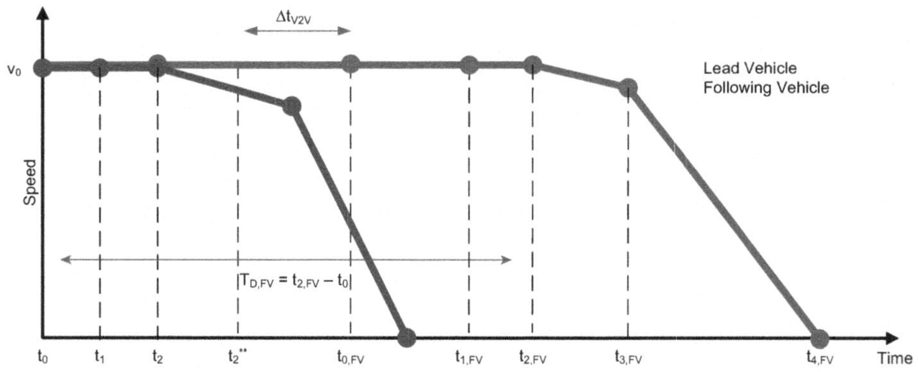

Fig. 8. Full braking, speed over time for leading and following vehicle (not true to scale)

The process of braking in the lead vehicle was already explained. Now the brake request of the lead vehicle needs to be transmitted to the follower. At t_2^{**} the brake request is transmitted over the air. The additional time lag between t_2 and t_2^{**} is due to data processing in the transmitting radio ECU for instance (e.g. signing, encryption). At $t_{0,FV}$ the brake request is received by the platoon follower. During dead time $T_{D,FV}$ between $t_{1,FV}$ and $t_{0,FV}$ additional steps compared to the leader need to be executed. The received data has to be verified, decrypted, transmitted from the radio to the Platooning ECU and finally transmitted to the brake system controller. After the reception of the brake request by the EBS at $t_{1,FV}$ the process is similar to the leading vehicle.

Now it is possible to set the respective time lags, initial speed and deceleration capabilities for the lead and following vehicle and calculate the overall stopping distance. This only delivers one result for a set of constant assumptions but in reality all of these dependencies vary in a certain range. To get a significant result it would be necessary to calculate all possible combinations within the whole range. This is where the Monte Carlo simulation comes into play.

3.3 Monte Carlo Simulations

3.3.1 Definition

Monte Carlo simulation got its name from Monte Carlo, Monaco. Casinos offering games of chance are the primary attractions. In games like roulette or roll the dice the outcome is random. You know the valid range of values, for roulette for instance you know that the wheel will show a number between 0 and 36 but you do not know the number for a particular trial. It is the same with variables of an equation. You know the range, the distribution and the certainty of the values but you do not know which value is randomly set for the actual situation. For example you know that Platooning will take place mainly on motorways at speeds between 70 and 90 kph for instance and that a most likely speed which is driven by the drivers is 85 kph but you do not know for any particular case.

For a Monte Carlo simulation (MCS) assumptions and a range of values have to be defined for every variable of interest. The simulator will generate random numbers based on these assumptions and distributions and will repeat the calculation of the equation. In this case 10,000 simulations have been done. A graphical outcome is generated which provides a certainty for a predefined event, in this case the difference of the overall stopping distance of platoon leader and follower. The used distributions are shown in Fig. 9.

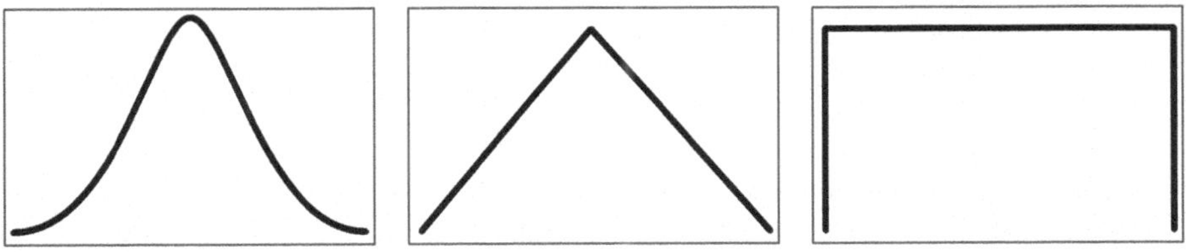

Fig. 9. Used variable distributions for Monte Carlo simulation: Normal, Triangular and Uniform distribution

3.3.2 Assumptions

As stated above the overall stopping distance depends on many variables for which assumptions have to be made as input for the Monte Carlo simulation. The assumed distribution as well as the range for the lead vehicle is contained in Table 1.

For the following vehicle the assumptions are contained in Table 2. Basically the values are identical since no separate preconditions – an improved brake performance for example – have been defined for the trailing truck in the first step. Keep in mind

Table 1. Monte Carlo variable assumptions for the lead vehicle

Variable	Explanation	Distribution	Unit	Min	Max	Mean/ Likeliest	Std Dev.
$t^*_{1,LV}$	Time at which full brake request (driver/autom. system) is received by the EBS	Uniform	[ms]	15	35		
$t_{1,LV}$	Data processing in EBS (calculation of pressure etc.)	Uniform	[ms]	4	6		
$t_{2,LV}$	Start of brake rising time (axle modulator receives pressure)	Triangular	[ms]	5	30	10	
$t^*_{2,LV}$	Data processing in Platooning ECU	Uniform	[ms]	4	6		
$t^{**}_{2,LV}$	Data processing in V2V radio (signing, encryption) and transmitting	Uniform	[ms]	4	6		
$t_{3,LV}$	End of brake rising time (full brake pressure applied in all brake chambers)	Triangular	[ms]	200	600[a]	350	
$v_{0,LV}$	Initial speed	Triangular	[km/h]	74	96[b]	85	
$a_{min,LV}$	Maximum braking capability (refers to [7])	Normal	[m/s²]	3.4	8.1	5.75	0.91
$k_{R,LV}$	Factor to determine the deceleration during brake rise time	Uniform	[-]	0.35	0.75		

[a] 600 ms is allowed maximum rising time according to UNECE R-13.
[b] Allowed tachograph tolerance 6 km/h (165/2014/EC); Speed limiter of N3 vehicles 90 km/h (92/6/EEC).

that the distribution and the range of values are identical but during the simulation different and independent values will be set randomly for the platoon leader and follower. However one precondition has been made: The initial condition is a stable following with an active automatically controlled distance, i.e. the initial speeds of both trucks are nearly identical.

One additional simulation without automatic Platooning controller was done as a reference. The following distance is set to 2 s as mandated or applied by most EU states. As reaction time for the driver of the following truck the "Kölner Modell" [6] was used,

Table 2. Monte Carlo variable assumptions for the following vehicle

Variable	Explanation	Distribution	Unit	Min	Max	Mean/ Likeliest	Std Dev.
$t_{0,FV}$	Reception of brake signal via V2V	Triangular	[ms]	30	300	100	
$t^*_{0,FV}$	Verifying and decryption of data	Uniform	[ms]	4	6		
$t^{**}_{0,FV}$	Data processing in Platooning ECU	Uniform	[ms]	4	6		
$t^{***}_{0,FV}$	Transmission of brake request to EBS	Triangular	[ms]	10	60	20	
$t_{0,FV,Drv}$	Driver reaction time	Triangular	[s]	0.78	1.49	1.25	
Δdx	Measured distance accuracy of RADAR	Uniform	[m]	−0.25	0.25		

that is 0.78 s as minimum, 1.49 s as maximum and 1.25 s as likeliest driver reaction time. A triangular distribution was applied. All other assumptions remained unchanged.

3.3.3 Results

Figure 10 shows the simulation result for a manually initiated braking in the following vehicle based on the driver reaction times mentioned above. The certainty of 6% for rear-end collisions means that even if the driver is driving at the legally mandated time gap of 2 s the crash cannot be avoided by all means. This result shall be used as a reference and shall represent the situation on public roads based on the legal mandate. Table 3 shows the certainties of all simulated time gaps.

The results for an automatic Platooning system in which the platoon members are electronically coupled are shown in Table 4. It can be seen, that under the assumptions made the certainty for rear end collisions below 1 s headway is higher than a manually driven truck keeping a 2 s following distance. This means, that a fast radio communication alone is not the solution for a safe platoon. For a following distance of 1 s the certainty is 5.7% and therefore matches the 6% of the manually driven truck at 2 s. The conclusion of this is that a platoon coupled by V2V is not safer than a manually driven truck if the following distance is shorter than 1 s and no other technical measures are taken.

Table 3. MCS results for manually driven following vehicle

Headway	[s]	2.0	1.8	1.4	1.2	1.0	0.8	0.5	0.3
	[m]	47.2	42.5	33.1	28.3	23.6	18.9	11.8	7.1
Certainty	[%]	6.0	10.9	34.6	50.8	66.8	79.7	91.8	96.1

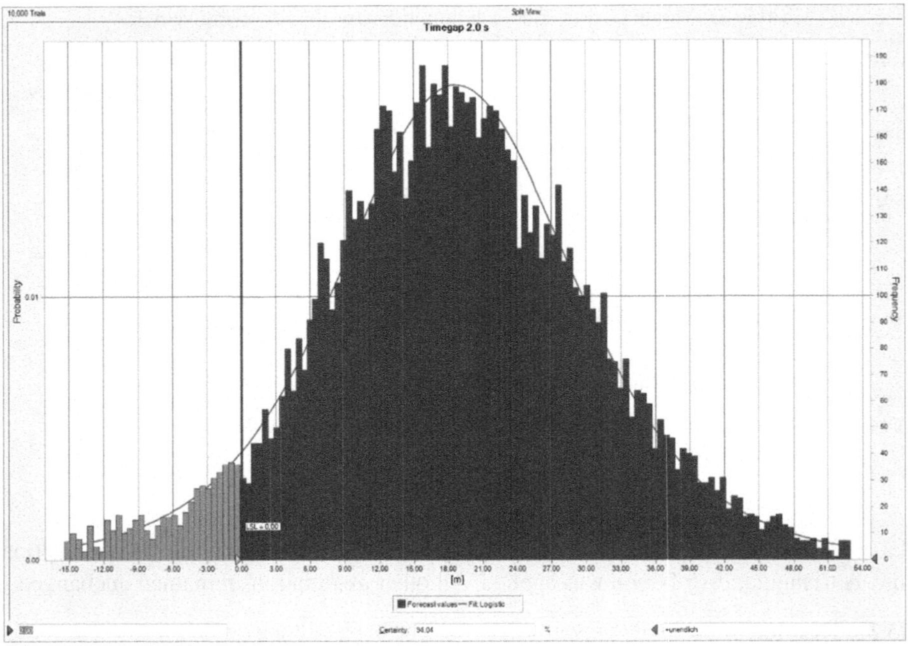

Fig. 10. MCS result for manually driven FV; time gap 2.0 s (certainty of rear-end collision: 6%)

Table 4. MCS results for a V2V based automated Platooning system

Headway	[s]	2.0	1.8	1.4	1.2	1.0	0.8	0.5	0.3
	[m]	47.2	42.5	33.1	28.3	23.6	18.9	11.8	7.1
Certainty	[%]	0.0	0.2	1.1	2.7	5.7	11.3	26.9	42.3

The main influence for the certainty of collisions is the deceleration capability. Reducing the time lags of V2V or the cycle time of an ECU will not change the situation considerably. If Platooning shall be safe the difference in brake performance must be reduced. It is not possible to determine the overall maximum brake performance of a truck without performing a full braking under the respective condition. So it will not be possible to completely exclude trucks which brake performance is for example less than 4.5 or 5 m/s^2. Simply because you just don't know (refer to influencing factors Sect. 3.1).

Possibilities to reduce the variance in distribution of the brake performance could be a better maintenance or shorter time interval in between. Installing a higher brake performance in the trucks and trailers, like bigger brake chambers, higher supply pressure, linings with high and constant friction coefficient. Here might be the trade off. Commercial vehicles need to earn money and with help of Platooning money shall be saved. If this has to be invested in a better brake performance it might eat up the savings (Fig. 11).

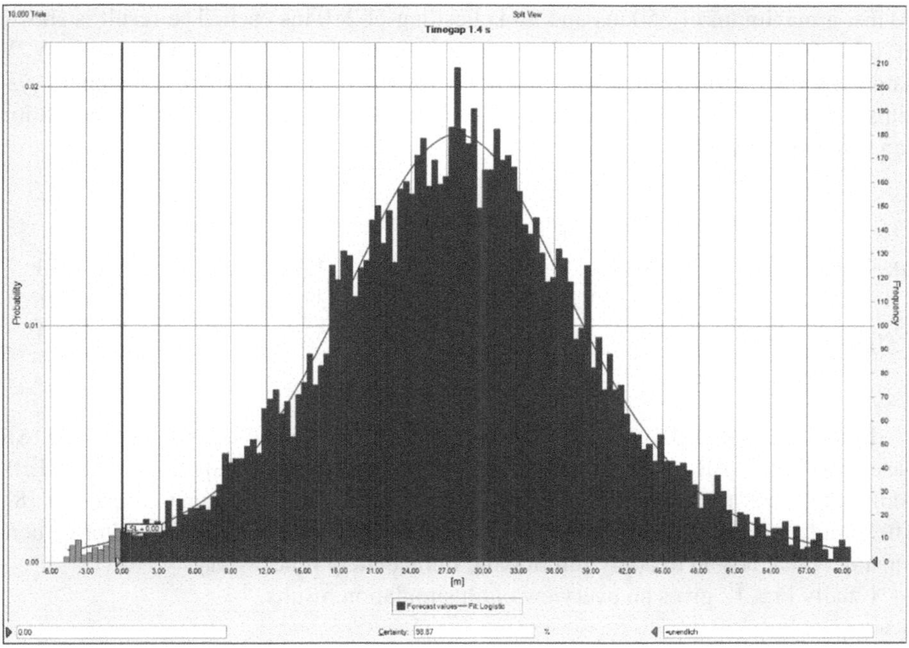

Fig. 11. MCS result for automated Platooning FV; time gap 1.4 s (certainty of rear-end collision: 1.1%)

To see the effect of a smaller variance an additional simulation was done with a reduced standard deviation for the maximum braking capability. All other factors remained constant. The standard deviation was reduced from 0.91 to 0.7. Table 5 contains the result.

Table 5. MCS results for an automated Platooning system with reduced standard deviation of 0.7

Headway	[s]	2.0	1.8	1.4	1.2	1.0	0.8	0.5	0.3
	[m]	47.2	42.5	33.1	28.3	23.6	18.9	11.8	7.1
Certainty	[%]	0.0	0.0	0.3	0.8	2.4	6.6	22.1	40.5

Compared to the unmodified simulation (Table 4) the collision certainties moved by 0.2 s. This means that the following distance can be reduced by 0.2 s for similar certainty. Precondition as stated would be a better brake performance and a smaller variance.

As a final variant a simulation without V2V but RADAR based only was conducted. Similar to the driver the RADAR based system can only react on the movement of the preceding vehicle but the RADAR is still faster than a human. Sensing and object tracking are assumed as triangular distribution with a minimum time lag of

70 ms, a maximum of 350 ms and a likeliest lag of 210 ms each. The result is shown in Table 6. It is foreseen that a Platooning system which mainly builds up on V2V data exchange needs a redundant system for safety reasons. RADAR and camera systems which are installed in today's trucks anyway due to AEBS and LDWS legislation can take over this role.

Table 6. MCS results for a RADAR based distance control

Headway	[s]	2.0	1.8	1.4	1.2	1.0	0.8	0.5	0.3
	[m]	47.2	42.5	33.1	28.3	23.6	18.9	11.8	7.1
Certainty	[%]	0.3	0.7	4.1	8.2	15.0	26.5	50.4	67.3

It shall be noted that such a system cannot be compared with today's RADAR based ACC or AEB systems which only work in certain system limits. An AEBS additionally has to follow a mandated cascade before full braking is requested [8]. The simulation refers to a modified distance control based on RADAR and/or camera sensing which is able to apply full braking immediately if necessary.

Finally Fig. 12 gives an overview of all simulation results.

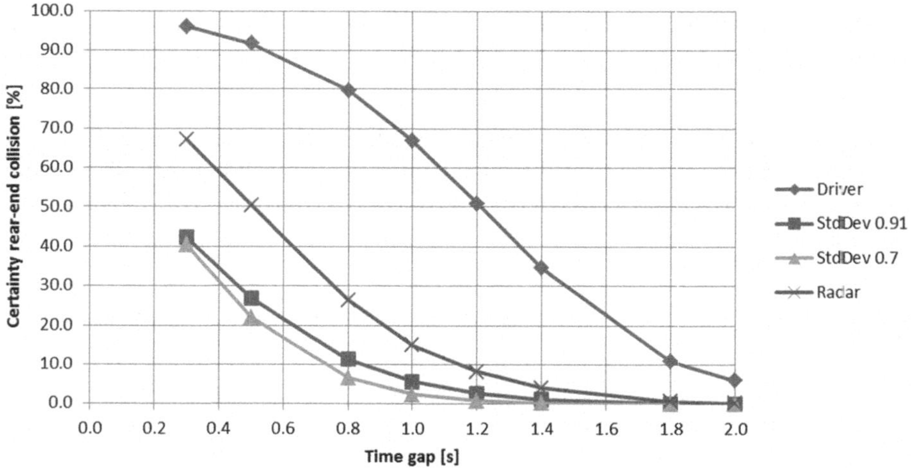

Fig. 12. Rear-end collision certainty dependent on time gap for conducted simulations

3.4 Proposal of a "Safe" Headway

The challenge always is to define "safe". Shall it mean that the automatic system is just as good as a human being or shall it be better? Based on the simulations a following distance of 1 s with V2V has a similar certainty for a rear-end collision as a driver with 2 s headway. No modifications or improvements done on truck and trailer brakes. It is at least questionable if authorities would follow such an argumentation or if it will be required that an automatic system should be better than a human driver. Assuming that a certainty of 1% is a value which will be followed a headway of 1.4 s could be considered as safe. If trucks and trailers which are allowed to platoon get a better brake performance and maintenance, that is reducing the variance in overall brake performance, a safe headway could be 1.2 or maybe 1 s. This statement is just based on the certainty of rear-end collisions. Regarding the fuel savings it does not matter if the following distance is 1, 1.2 or 1.4 s (compare Sect. 2).

4 Platooning Concept Proposal

Currently all major truck OEMs are working on Platooning systems which aim for distances less or equal than 0.8 s to gain maximum possible savings. Up to now all of these are proprietary and not compatible among manufacturers. Questions like common standards on the communication, differences in brake performance; legal hurdles etc. are not solved yet. A European funded project called ENSEMBLE[3] – at which 6 major EU truck OEMs participate – shall address these topics.

Instead of waiting for such highly automated systems a pre stage which addresses two of the three mentioned goals for Platooning (see Sect. 1.2) – that is reducing emissions and fuel usage and increase road safety – could be realized easier and faster.

Regarding road safety today we have the situation that truck drivers often do not respect the mandated headway and follow each other too closely. AEBS which was introduced for all new trucks licensed after November 2015 can be switched off and many drivers seem to use this option. Even with AEBS enabled the collision might not be avoided due to the warning cascade or the too close distance. A better option would be a continuous distance control like ACC. But as a comfort system ACC only works in certain limits, the maximum requestable deceleration is limited for instance. Also it was developed for the legal headway of 2 s. To save fuel it uses up the intervehicle distance in certain limits to avoid too much braking.

A solution could be a modified ACC and AEBS which legalizes following distances down to 1.4 or 1.2 s which are driven anyway. Such a system could reduce the number of rear-end collisions. The existing ACC and AEBS systems need to be modified of course. They need to react faster and stronger than current systems. This application modification is one side. On the other hand there are still the system limits for RADAR systems and therefore limitations on the applications using this input. Known problems are the false positives due to object tracking and misclassification of stationary objects. If the truck is not in platooning mode, all of these filters and limitations shall remain active, but in platooning mode, the RADAR has only one moving object in front which is relatively close. The occurrence probability of a misclassification is quite low which means that in these conditions ACC and AEB system can put more trust in the provided signals. Such a system should be designed in a way, that the driver cannot switch it off or with other words: If the driver wants to drive closer than

[3]ENSEMBLE shall be startet in 2018 and shall last 3 years.

2 s he has to enable this new system. The current mandated AEBS of course should be enabled all the time.

Following distances of 1.2 to 1.4 s already would save a noticeable amount of fuel as demonstrated in Sect. 2. Putting a system in place which has an adapted ACC and AEBS algorithm and therefore realizes a smaller following distance could be a first step. Such a system might also get a higher acceptance amongst the other traffic participants. The headways are not too close, questions like opening up a platoon in tunnels, on bridges or in front of entry and exit ramps do not have to be asked. It would lead to a convoy in which basically everybody is follower and can save fuel.

It is worthwhile pointing out that independent to Platooning available aero dynamic kits (like OptiFlow™) achieve fuel savings in the same magnitude as Platooning is promising.

A Appendix

A.1 SAE J3016 – Levels of Driving Automation

See Fig. 13.

SAE level	Name	Narrative Definition	Execution of Steering and Acceleration/ Deceleration	*Monitoring* of Driving Environment	Fallback Performance of *Dynamic Driving Task*	System Capability (*Driving Modes*)
Human driver **monitors the driving environment**						
0	No Automation	the full-time performance by the *human driver* of all aspects of the *dynamic driving task*, even when enhanced by warning or intervention systems	Human driver	Human driver	Human driver	n/a
1	Driver Assistance	the *driving mode*-specific execution by a driver assistance system of either steering or acceleration/deceleration using information about the driving environment and with the expectation that the *human driver* perform all remaining aspects of the *dynamic driving task*	Human driver and system	Human driver	Human driver	Some driving modes
2	Partial Automation	the *driving mode*-specific execution by one or more driver assistance systems of both steering and acceleration/ deceleration using information about the driving environment and with the expectation that the *human driver* perform all remaining aspects of the *dynamic driving task*	**System**	Human driver	Human driver	Some driving modes
Automated driving system **("system") monitors the driving environment**						
3	Conditional Automation	the *driving mode*-specific performance by an *automated driving system* of all aspects of the dynamic driving task with the expectation that the *human driver* will respond appropriately to a *request to intervene*	System	**System**	Human driver	Some driving modes
4	High Automation	the *driving mode*-specific performance by an automated driving system of all aspects of the *dynamic driving task*, even if a *human driver* does not respond appropriately to a *request to intervene*	System	System	**System**	Some driving modes
5	Full Automation	the full-time performance by an *automated driving system* of all aspects of the *dynamic driving task* under all roadway and environmental conditions that can be managed by a *human driver*	System	System	System	**All driving modes**

Fig. 13. SAE J3016 levels of driving automation [9]

A.2 Table Fuel Savings North American Vehicles

Table 7. Platooning fuel savings (North American vehicles)

	Year	location		Speed [mph]	Speed [kph]	Distance [ft]	Distance [m]	Distance [s]	Fuel savings [%] LV	Fuel savings [%] FV	load [lbs]	load [kg]	aero fitted
PATH 50..55 mph (80..89 kph)	2003	Testtrack	no traffic	50..55	80..89	10	3	0.1	9.2	11.6			
						13	4	0.2	9.1	12.2			
						20	6	0.2	7.1	9.2			
						26	8	0.3	5.6	10.8			
						33	10	0.4	6.0	10.0			
Peloton 64 mph (103 kph)	2013	Highway	minimal traffic	64	103	36	11	0.4	4.5	10.0			
NREL SAE 55..70 mph (89..113 kph)	2014	Testtrack	no traffic	55	89	30	9	0.4	4.3	8.4			
				55	89	50	15	0.6	2.2	9.7			
				65	105	20	6	0.2	5.3	2.8	engine fan		
				65	105	30	9	0.3	4.1	7.5			
				65	105	40	12	0.4	2.7	9.1			
				65	105	50	15	0.5	3.1	9.2			
				65	105	75	23	0.8	1.7	9.4			
				70	113	30	9	0.3	4.4	4.6	engine fan		
Peterbilt / EXA / [illegible] 60 mph (97 kph)	2014	Simulation	no traffic	60	97	30	9	0.3	3.8	9.2			
						60	18	0.7	0.5	10.8			
						100	30	1.1	0.0	9.2			
						160	49	1.8	0.0	7.6			
	2017					316.8	97	3.6	0.0	3.2			
NRC Canada 55/65 mph (89/105 kph)	2017	Testtrack	no traffic	65	105	57	17	0.6	1	12.3	65000	29400	yes
						86	26	0.9	0.3	11.7			
						114	35	1.2	-0.4	10.4			
						143	44	1.5	0.3	10.4			
				65	105	57	17	0.6	0.3	11.0	65000	29400	no
						86	26	0.9	-0.7	9.9			
						114	35	1.2	-0.4	9.8			
						143	44	1.5	0	9.5			
				55	89	56	17	0.7	1.6	10.5	65000	29400	no
				65	105	57	17	0.6	1.4	12.1	31000	14000	no
				65	105	57	17	0.6	8.5	17.9	65000	29400	yes
						86	26	0.9	7.9	17.3			
						114	35	1.2	7.2	16.1			
						143	44	1.5	7.9	16.0			
[illegible] NA 65 mph (105 kph)	2017	Testtrack TRC Ohio	no traffic	65	105	134	41	1.4	0.9	14.8	65000	29400	yes
						343	105	3.6	0.9	12.6			
				65	105	43	13	0.5	3.3	7.1			no
						86	26	0.9	1.6	7.1			
						134	41	1.4	1.5	5.7			
						343	105	3.6	1.9	3.0			

A.3 Table Fuel Savings Cabovers

See Table 8.

Table 8. Platooning fuel savings (cabovers)

	Year	location		Speed [mph]	[kph]	Distance [ft]	[m]	[s]	Fuel savings [%] LV	FV
CHAUFFEUR 50 mph (80 kph)	2000	Testtrack	no traffic	50	80	22	6.7	0.27	9.5	20.4
						23	7.0	0.28	8.5	20.6
						26	7.9	0.32	8.5	21.0
						33	10.1	0.41	6.0	20.8
						39	11.9	0.48	5.0	19.0
						46	14.0	0.57	3.0	17.3
						53	16.2	0.65	0.0	15.7
SARTRE / VOLVO 56 mph (90 kph)	2010 - 2012	Testtrack	no traffic	56	90	16	5	0.20	7.5	16
						20	6	0.24	8.4	14.0
						23	7	0.28	7.5	12.5
						26	8	0.32	7.0	12.1
						30	9	0.36	5.4	10.0
						33	10	0.40	5.2	9.9
						39	12	0.48	3.7	8.1
						49	15	0.60	1.1	8.1
Japan Energy ITS 50 mph (80 kph)	2013	Testtrack	no traffic	50	80	16	5	0.2	9.0	15.7
						33	10	0.4	3.3	16.5
						39	12	0.5	2.5	16.1
						49	15	0.6	1.0	16.1
						66	20	0.8	0	15
[illegible] EU-1 53 mph (85 kph)	2017 January	Testtrack	no traffic	53	85	33	10	0.4	3.9	10.1
						66	20	0.8	-0.3	8.6
						98	30	1.3	0.7	7.6
						131	40	1.7	0.7	7.8
						164	50	2.1	0.8	8.3
[illegible] EU-2 53 mph (85 kph)	2017 August	Testtrack	no traffic	53	85	33	10	0.4	2.8	7.2
						66	20	0.8	-0.1	8.4
						98	30	1.3	0.4	7.1
						131	40	1.7	-0.4	7.1
						164	50	2.1	-0.2	7.4
						230	70	3.0	-0.8	6.6

A.4 Following Distances in the USA

See Table 9.

Table 9. Following distances for HGV in the USA

State	Speed limit		Headway			"Safe distance" "reasonable and prudent"	Comments
	[mph]	[kph]	[m]	[ft]	[s]		
Alabama	70	113	91.44	300	2.9	x	
Alaska	65	105				x	
Arizona	75	121				x	
Arkansas	70	113	60.96	200	1.9	x	
California	55	89				x	
Colorado	75	121				x	
Connecticut	65	105				x	
Delaware	65	105	91.44	300	3.1	x	
District of Columbia	55	89					
Florida	70	113	91.44	300	2.9	x	
Georgia	70	113				x	
Hawaii	60	97				x	
Idaho	70	113				x	
Illinois	70	113				x	
Indiana	65	105					
Iowa	70	113				x	
Kansas	75	121				x	
Kentucky	70	113	76.2	250	2.4		
Louisiana	75	121			2.0	x	The Two Second Rule is recommended as reasonable and prudent by Louisiana's driver's license guide book
Maine	75	121				x	
Maryland	70	113				x	
Massachusetts	65	105	60.96	200	2.1		
Michigan	60	97	152.4	500	5.7	x	
Minnesota	70	113	152.4	500	4.9	x	
Mississippi	70	113	91.44	300	2.9	x	
Missouri	70	113				x	
Montana	65	105				x	

(Continued)

Table 9. *(Continued)*

State	Speed limit		Headway			"Safe distance" "reasonable and prudent"	Comments
	[mph]	[kph]	[m]	[ft]	[s]		
Nebraska	75	121	30.48	100	0.9	x	
Nevada	80	129	152.4	500	4.3	x	
New Hampshire	70	113				x	
New Jersey	65	105	30.48	100	1.0		100 ft is the threshold to commit the offense "tailgating"
New Mexico	75	121	91.44	300	2.7		
New York	65	105				x	
North Carolina	70	113				x	
North Dakota	75	121				x	
Ohio	70	113				x	
Oklahoma	75	121				x	
Oregon	65	105				x	
Pennsylvania	70	113				x	
Rhode Island	65	105				x	
South Carolina	70	113				x	
South Dakota	80	129				x	
Tennessee	70	113				x	
Texas	85	137				x	
Utah	80	129			2.0	x	
Vermont	65	105				x	
Virginia	70	113				x	
Washington	60	97				x	
West Virginia	70	113	60.96	200	1.9		
Wisconsin	70	113	152.4	500	4.9		
Wyoming	80	129				x	

A.5 Following Distances in Europe

Table 10. Following distances for HGV in the Europe

	Headway		"Safe distance"	Comments
	[m]	[s]		
Austria	50			2 s rule is taught in driving schools
Denmark		2		2 s rule is applied
Estonia			x	2 s rule is a rule of thumb
Finland			x	2 s rule is a rule of thumb
France	50			
Germany	50			
Iceland			x	
Ireland		2*	x	2 s rule is part of "Rules of the road" but not a legal requirement
Italy			x	
Luxembourg		2*	x	2 s rule is a rule of thumb
The Netherlands		2*		
Norway			x	Fines for HGV is distance is less than 1 s
Portugal			x	
Switzerland			x	2 s rule is applied in court
Sweden			x	Fines for HGV is distance is less than 1 s
UK		2*	x	2 s rule is part of "Rules of the road" but not a legal requirement

A.6 Metric/Imperial Units (Speed, Distance, Headway)

Table 11. Overview of metric/imperial speeds, distances and headways

Speed			
Metric		Imperial	
[kph]	[m/s]	[mph]	[ft/s]
1.0	0.3	0.6	0.9
3.6	1.0	2.2	3.3
5.0	1.4	3.1	4.6
10.0	2.8	6.2	9.1
15.0	4.2	9.3	13.7
16.1	4.5	10.0	14.7
20.0	5.6	12.4	18.2
25.0	6.9	15.5	22.8
30.0	8.3	18.6	27.3
32.2	8.9	20.0	29.3
40.0	11.1	24.9	36.5
50.0	13.9	31.1	45.6
60.0	16.7	37.3	54.7
70.0	19.4	43.5	63.8
80.0	22.2	49.7	72.9
85.0	23.6	52.8	77.5
89.0	24.7	55.3	81.1
90.0	25.0	55.9	82.0
96.6	26.8	60.0	88.0
100.0	27.8	62.1	91.1
104.6	29.1	65.0	95.3
112.7	31.3	70.0	102.7
120.7	33.5	75.0	110.0
128.7	35.8	80.0	117.3
136.8	38.0	85.0	124.7

Distance		
Metric	Imperial	
[m]	[ft]	[yd]
1.0	3.3	1.1
0.3	1.0	0.3
5.0	16.4	5.5
6.1	20.0	6.7
9.1	30.0	10.0
10.0	32.8	10.9
12.2	40.0	13.3
15.0	49.2	16.4
15.2	50.0	16.7
20.0	65.6	21.9
22.9	75.0	25.0
25.0	82.0	27.3
30.0	98.4	32.8
30.5	100.0	33.3
35.0	114.8	38.3
40.0	131.2	43.7
45.0	147.6	49.2
45.7	150.0	50.0
50.0	164.0	54.7
61.0	200.0	66.7
70.0	229.7	76.6
76.2	250.0	83.3
80.0	262.5	87.5
91.4	300.0	100.0
100.0	328.1	109.4

Headway (Following Distance)				
Time	Metric		Imperial	
[s]	[kph]	[m]	[mph]	[ft]
0.3	80.0	6.7	49.7	21.9
	90.0	7.5	55.9	24.6
	96.6	8.0	60.0	26.4
	104.6	8.7	65.0	28.6
	120.7	10.1	75.0	33.0
0.5	80.0	11.1	49.7	36.5
	90.0	12.5	55.9	41.0
	96.6	13.4	60.0	44.0
	104.6	14.5	65.0	47.7
	120.7	16.8	75.0	55.0
0.8	80.0	17.8	49.7	58.3
	90.0	20.0	55.9	65.6
	96.6	21.5	60.0	70.4
	104.6	23.2	65.0	76.3
	120.7	26.8	75.0	88.0
1.8	80.0	40.0	49.7	131.2
	90.0	45.0	55.9	147.6
	96.6	48.3	60.0	158.4
	104.6	52.3	65.0	171.6
	120.7	60.4	75.0	198.0
3.6	80.0	80.0	49.7	262.5
	90.0	90.0	55.9	295.3
	96.6	96.6	60.0	316.8
	104.6	104.6	65.0	343.2
	120.7	120.7	75.0	396.0

Abbreviations/Acronyms

ACC	Adaptive Cruise Control
AEBS	Advanced Emergency Braking System
COE	Cab-Over-Engine Trucks/also referred as cabover trucks or cabovers
CFD	Computational Fluid Dynamics
CAN	Controller Area Network
C-ITS	Cooperative ITS
EBS	Electronic Braking System
ESC	Electronic Stability Control
ECU	Electronical Control Unit
EU	Europe
FV	Following Vehicle (Platoon follower)
HCV	Heavy Commercial Vehicle
HGV	Heavy Goods Vehicle
IEEE	Institute of Electrical and Electronics Engineers
ITS	Intelligent Transport Systems
LDWS	Lane Departure Warning System
LV	Lead Vehicle (Platoon leader)
MCS	Monte Carlo Simulation

NREL	National Renewable Energy Laboratory (https://www.nrel.gov/)
NRC	National Research Council Canada (https://www.nrc-cnrc.gc.ca/)
NA	North America(n)
NACFE	North American Council for Freight Efficiency (https://nacfe.org/)
OEM	Original Equipment Manufacturer
CHAUFFEUR	PROMOTE CHAFFEUR II, EU funded research project ended May, 2003
RADAR	Radio Detection and Ranging
RDW	Rijksdienst voor het wegverkeer (Netherlands Vehicle Authority)
SARTE	Safe Road Trains for the Environment – EU funded research project ended Oct, 2012
SAE	Society of Automotive Engineers
STD	Standard trailer configuration, i.e. no aerodynamic devices installed
AERO	Trailer equipped with aerodynamic devices like side-skirts and boat-tails
TRC	Transportation Research Center Inc. (http://www.trcpg.com)
UNECE	United Nations Economic Commission for Europe (https://www.unece.org)
USA	United States of America
V2V	Vehicle to Vehicle communication based on Wi-Fi-p (IEEE 802.11p)
Wi-Fi	Wireless local area networking based on IEEE 802.11 standards

References

1. NRC (National Research Council Canada): Fuel-economy testing of a three-vehicle truck platooning system (2017)
2. NACFE (North American Council for Freight Efficiency): CONFIDENCE REPORT: Two-Truck Platooning (2016)
3. Peterbilt, EXA Corporation: Aerodynamic impact of tractor-trailer in drafting configuration (2014-01-2436). In: SAE International (2014)
4. NREL (National Renewable Energy Laboratory): Effect of Platooning on Fuel Consumption of Class 8 Vehicles Over a Range of Speeds, Following Distances, and Mass (2014-01-2438). In: SAE International, 30 September 2014
5. SAE, J1321 – Fuel consumption test procedure – Type II, 2012-02
6. Colliseum: Kölner Modell. https://www.colliseum.net/wiki/K%C3%B6lner_Modell. Accessed 5 Mar 2018
7. Investigation of brakes of heavy vehicles. Færdselsstyrelsen (2005)
8. 347/2012/EC, 2012-04-16
9. SAE, J3016 – Taxonomy and definitions for terms related to driving automation systems, 2016-09

aFAS – How to Get a Driverless Prototype on the Road?

Patrick Jiskra(✉), Peter Strauß, and Walter Schwertberger

MAN Truck & Bus AG, München, Germany
patrick.jiskra@man.eu

1 aFAS – How to Get a Driverless Prototype on the Road?

Welcoming sentence

Mobile road works on highway hard shoulder lanes are usually protected against on- coming traffic by protective vehicles equipped with a mobile warning trailer. The service staff operating the protection vehicle for road maintenance works have a high degree of risk of being involved into severe accidents with surrounding traffic. With the motivation to reduce this risk, the "Automatic unmanned protective vehicle for mobile construction sites on motor highways" (aFAS) project, funded by the Federal Ministry for Economic Affairs and Energy (BMWi), started in August 2014. The aim of this project is to develop an unmanned protective vehicle which is capable to perform this task driverless by automatically following mobile roadworks – or a preceding work vehicle equipped with V2V technology respectively.

This presentation presents the basic objectives of the aFAS project and gives an insight into the system realization. The first use of a unmanned protective vehicle on public roads presents special challenges in the functional safety of driverless operation. Based on the ISO 26262 standard, the elaborated safety concept is presented and linked with the technical concept and the resulting test concept.

2 Agenda

Short sentences about each topic and subtopic.

3 Project aFAS

Automatisch fahrerlos fahrendes Absicherungsfahrzeug für Arbeitsstellen auf Autobahnen.

"Automatic unmanned protective vehicle for mobile construction sites on motor highways" (aFAS) is a project funded by the Federal Ministry for Economic Affairs and Energy (BMWi) since August 2014.

T. Bertram (Hrsg.): *Fahrerassistenzsysteme 2018,* Proceedings, S. 158–165, 2019.
https://doi.org/10.1007/978-3-658-23751-6_15

A consortium of eight partners from industry, research and administration is developing this prototype of protective vehicle. An protective vehicle is a vehicle that follows at a typical distance of 50 to 100 m a work vehicle while this is working on the side of a motorway. The typical tasks are to warn surrounding traffic and to protect the crew/workers against collisions with oncoming vehicles.

The aim of this project is to develop a SAE Level 4 driverless and fully automated vehicle on public roads. This implies the design of a comprehensive safety concept as well as the administrative and legal implementation.

3.1 Operation and Service

Road Maintenance Works. Road service staff work at regular road maintenance works on motor highways such as cutting of overgrown grass, bushes and trees, but also at road sweeping, litter collection, and cleaning of signs, reflectors and drains.

Operating Conditions. These road maintenance works are usually carried out in predefined, specified distances i.e. in between two highway access nodes/lanes. They are done at a slow working speed of about 2 to 10 kph. Road sweeping and grass mowing for example is done at an average speed of 3 kph.

Motivation and Goals. Mobile road works on the side of highways are usually protected against traffic by an protective vehicle fitted with a warning trailer. The service staff working on these shorttermed road maintenance jobs are at risk from heavy accidents, despite extensive safety precautions. This applies in particular to the crew of the security vehicle, which secures the workplace backwards against the traffic. Manned protective vehicles with warning trailers in use are regularly involved in heavy traffic accidents in which mainly employees of the road service come to harm.

More than 100,000 short duration road works are set up and secured throughout Germany every year. The probability for road service staff for being involved in an accident is statistically "about […] 13 times higher than [for employees] in the commercial sector" (Bruchsaler Rundschau 2017). Approximately two thirds of the accidents in AkD are caused by trucks. Mostly these are side or rear impacts. 80% of injuries or fatalities arise in scenarios where the crews are located inside their vehicles. In more than 63% of externally-inflicted accidents with vehicles of the road service on federal motorways protective vehicles/equipment are affected. There is also an increased risk for employees to be involved in accidents while entering or leaving their vehicles or setting up belay devices (Bast 2006).

By automated operation of such an protective vehicle this threat can be eliminated, at least on the part of road service staff.

3.2 The Vision for Higher Security

The advantages of protecting mobile short-term construction sites on highway shoulders with a unmanned protective vehicle are:

- No personnel in the danger zone (protective vehicle)
- No coupling or decoupling in the danger zone
- Frequency of leaving/entering vehicles on the highway is highly reduced – one operator has to switch vehicles only at start and end of the operation

Partners. In addition to industry representatives (MAN Truck & Bus AG (consortium leader), ZF TRW, WABCO Development GmbH, Robert Bosch Automotive Steering GmbH), the project consortium consists of representatives from science and research (Karlsruhe University of Technology, Technical University of Braunschweig) and the end-users the Road and Traffic Management Hessen Mobil and the Federal highway research institute (BASt).

Project Goals. The main project goal is the proof of feasibility of a fully automated protective vehicle in a very simple scenario. The final phase of the project consists in a pilot operation on Hessian highways embedded in real operation in 2018 on a highway of traffic in Hessen.

3.3 Functionality

This slide gives an overview on how the system works.

The potential use of the vehicle is on the shoulder lane for mobile and stationary work sites and as an early warning vehicle.

It shall protect mobile short-term construction sites on highway shoulders.

Scenario. A typical deployment scenario for the unmanned protective vehicle is as follows: At the beginning of the mission, working and protected vehicle have to be transported to the road works site. Both vehicles are operated manually until this point. At the workplace, the protective vehicle is parked at the hard shoulder behind the preceding working vehicle. Before switching to the working vehicle, the driver (engages the parking switch and driving position of the automatic gearbox) activates the automatic driving mode via user interface. In the working vehicle, the driver activates the driverless operation of the protective vehicle via a user interface. The vehicle guidance system takes over the longitudinal and lateral guidance of the security vehicle and follows the leader vehicle at a sufficiently large distance at a maximum operating velocity of 12 kph. The reference for the lateral control is the hard shoulder lane determined by detected the marking line. For the longitudinal control, the reference is the tracking of the preceding vehicle (free space detection included).

Driverless Modes. The process depicted describes one of three modes, called follow mode where the protective vehicle follows the work vehicle at a distance of 100 m. This is the main operating mode to be used during the actual works. This distance is chosen as mobile road works in Hessen are secured with this particularly distance based on Hessian road maintenance guidelines, but can values vary between different federal states.

Passing over access and exit lanes of motorways is also part of the functional scope of the driverless operation of the security vehicle. The safety vehicle follows the lead vehicle at a small distance to prevent surrounding traffic from intruding into the space between the vehicles. The Lateral control reference is now the target lateral offset. This is requested separately from the leading vehicle, whereupon the protective vehicle is closing up to the lead vehicle. On the one hand, this ensures that the mobile road work site safely passes the access or exit lanes and does not pose an obstacle to flowing traffic. On the other hand, in order to reduce risk, the intrusion between the leading vehicle and the driverless security vehicle is made more difficult by their short distance. After automated use, the driver changes back to the security vehicle and drives it back to the depot manually.

The last mode is the safety stop at a defined position which can either be selected by the driver of the work vehicle or is used in case of overpassing system boundaries.

4 Functional Safety According to ISO 26262

4.1 How Do We Get a Driverless Prototype on the Road?

As the protective vehicle has to be capable to operate completely driverless during operation at the hard shoulder, the designed system fundamentally differs from automated driving functions currently available in road vehicles, such as e.g. Adaptive cruise control systems, lane keeping systems or automatic parking systems. In these systems, the driver serves as a fallback for the event of a fault or for functional gaps. His task is to permanently monitor the condition of the vehicle – or control the automation system respectively- and, if necessary, to take over the vehicle management in critical driving situations.

The monitoring of driverless operation by the driver is not applicable in the aFAS project. Also, permanent monitoring from the work vehicle, hereafter referred to as the lead vehicle, is not sufficient because it is assumed that the supervisor is unable to guarantee permanent monitoring.

The electronic vehicle guidance system in the security vehicle is therefore according to the definitions of the SAE classified as fully automated or of high-automation. The use of highly or fully automated vehicles on public roads requires a permission of the road traffic law.

To carry out the automated driving task safely – from a functional safety perspective – the system has to cover the surveillance task formerly assigned to a human driver and has to be able to monitor its own state and system boundaries and to autonomously restore a safe state if errors occur.

As already mentioned, the goal of this project is to demonstrate a proof of feasibility for automated prototype vehicles in public traffic, which includes more specifically a proof of feasibility of protection, self-monitoring and independent safe degradation.

Parallel to this, the use on the side of a motorway and the low working speeds of the work stations of shorter duration, in contrast to the complex traffic, e.g. in cities comparatively simple boundary conditions. This affects both the detection and interpretation of the vehicle environment as well as the behavior in case of error. Therefore, the described application is suitable for the first implementation of a vehicle guidance system for driverless vehicles on public roads. Particular attention is paid to the development of a safety concept.

For the series development of electronic systems in road vehicles, the ISO 26262 standard reflects the state of the art in terms of functional safety. Although the project described here deals with a prototype implementation of a unmanned protective vehicle, a safety concept according to the ISO 26262 standard is to be developed. This is aimed at two aspects. First, a safety concept is to be defined, which can serve as a basis for later series development. Second, the basic applicability of the ISO standard to fully automated vehicle systems will be examined at the same time.

From this, five main objectives for the project aFAS result:

- Development of an automatically driverless safety vehicle for workplaces on motorways
- Development of a safety concept according to the standard ISO 26262 and its implementation in the protection vehicle
- Development of a technical safety concept, the testing and release procedure
- Consideration of legal aspects of automated driving and limits of relevant standards of functional safety, in particular ISO 26262
- First use of a driverless vehicle on public roads

Against this background, a consortium of eight partners from industry, research and administration is developing the prototype of a security vehicle, which fully automatically follows a work vehicle while working on the side of a motorway and protects against flowing traffic. The aim is also the first use of a driverless and fully automated moving vehicle on public roads. The design of a comprehensive security concept as well as the administrative and legal implementation, which are developed in this project as an example for fully automated driverless driving, are also necessary here.

Consortium including BASt, TU Braunschweig, HS Karlsruhe.

4.2 Item Definition and Hierarchy Diagram

In contrast to series development, only the concept phase of the ISO 26262 is performed. Thus we performed an item definition, HARA, FuSaCo and TeSaCo. This safety frame shall reduce the efforts for the prototypical system while keeping a persistent safety standard.

Before starting with a hazard assessment and risk analysis, the items have to be defined. Within such an item definition, first assumptions on the architecture are set. In the concept phase of the ISO, the system is divided into functional parts with relations, dependencies and interfaces.

Error propagation difficult, dependencies not visible, so hierarchy diagram using skill graphs to represent all functionalities. Position in hierarchy depends on level of abstraction, dependencies to functional elements visible. Easy detection of error propagation for HARA.

4.3 Hazard Assessment and Risk Analysis

The aim of a Hazard Assessment and Risk Analysis is to identify and classify hazards from malfunctions.

First step to identify hazardous events which are basically situations with a hazard that arises due to a function malfunction. Second step is to evaluate this hazardous event and classify it in Automotive Safety Integrity Levels (ASIL). Smallest level is QM (Quality Management), then ASIL A to D. The classification is categorized in exposure, severity and controllability which together form the ASIL. The possible combinations and resulting ASILs can be seen in the table.

Next slide

Here you can see an extract of the HARA for our project aFAS. The most critical hazard is the intrusion of right driving lane. We're in the follow mode at a nominal distance of 100 m and follow the shoulder lane by detecting lane markings visually. In case of a malfunctioning perception or lateral guidance, the protective vehicle might intrude the right driving lane.

As usual scenario around Frankfurt, there is a convoy of trucks on the right driving lane. When the protective vehicle intrudes the right driving lane, the first truck might evade, but the following truck then hits the aFAS.

The ASIL classification is done in several Workshops with all partners and the TU Braunschweig (Experts in FuSa) as leader and later on reviewed by TÜV SÜD.

This hazard is classified ASIL D

The severity is classified as S3 (fatal injuries) because the collision will happen with a high differential speed. Classifying severity already shows one deficiency of the ISO as no material damage is considered.

The exposure is classified as E4 (high probability) as driving on the shoulder lane is the regular operation.

In case of the controllability, this hazard is classified as a C3 (difficult or not controllable) as the AFA might turn at a full wheel angle in the first driving lane.

Here we looked at the controllability of the hazard for the affected road users as no controllability can be defined for a driverless system.

For the following traffic it is difficult or impossible to detect this fast manoeuver and to perform an adequate avoidance and braking maneuver.

In the following a decomposition is done using the ASIL D control unit of the Steering System to limit the steering wheel angle and rate. This reduces the severity and controllability for traffic.

The severity is classified as S2 (Serious and life-threatening injuries (survival likely)) because the collision will happen with a high differential speed, but the trucks will only touch first, as the steering angle and rate are limited radically. Classifying severity already shows one deficiency of the ISO as no material damage is considered.

The exposure is classified as E4 (high probability) as driving on the shoulder lane is the regular operation.

In case of the controllability, this hazard is classified as a C2 (normally controllable). AFA drifts slowly (e.g., 0.4 m/s lateral) onto the right lane. The following traffic is able to detect this despite being hidden by the truck as well as to perform an avoidance and braking maneuver.

This hazard leads to the safety goal 12.

Extract of several safety goals identified from HARA.

4.4 Functional Safety Concept

Safety goals result in functional safety requirements.
Here exemplary SG12 is illustrated with the strategy to avoid this hazard.
This leads to different functional safety requirements for the components.
FMEAs are done by partners on a component level.

4.5 Technical Safety Concept

Functional safety requirements lead to technical safety concept and hence technical requirements for the components.

Specifically for SG12, the highlighted paths show the connected components.

In short, the colors define the ASIL classification, so that white = QM, green = ASIL A, yellow = ASIL B, orange = ASIL C, red = ASIL D, whereas the form distinguishes between oval = sensor, circle = actuator, angular = control unit, triangle = display, hexagon = control element.

Abbreviations are used for: MOT = engine, GET = transmission, BRK = service brake, HBR = parking brake, LEN = active steering.

4.6 Vehicle Architecture

Test Vehicle MAN TGM 18.340 4x2

- Usable series components
  - Existing E/E architecture
  - Interfaces to engine and automated manual transmission
  - Infotainment system MMT for HMI display, standard button for mode selection
  - Serial sensors and systems such as AEBS as additional redundancy
- Required extensions
  - Electrohydraulic steering actuator from Bosch
  - Fail-Operational Brake System from WABCO
  - Field survey of the company ZF TRW
  - Vehicle – Vehicle – Communication according to WLAN standard IEEE 802.11p

4.7 Testing, Release and Documentation

268 different test cases

Test Concept

Technical Requirements derived
Software Tests (based on ISO 29119)
Test drives to support the safety concept
Tests in real environment
Pilot operation in Hessen

Release

Release Process by experts of TU BS
Review by TÜV SÜD

Evaluation

From the perspective of the maintenance service influences on other road users
Legal evaluation of fully automated driverless vehicles
Proposal of guidelines to support the approval of the experimental operation on public roads

5 Summary

It is already possible today to send such a prototype to the street with reasonable risk. Even a partial selection of the ISO kit helps us. The safety framework concentrates testing and protection effort on a part of the system that is even partly reusable. Experiences and findings from this project will flow into many more autonomous driving projects. The problem that the current jurisprudence is still not ready for driverless driving in public transport, plays a minor role in prototypes, as special permits can be granted.

Important from an OEM perspective is the structure of the project: Partnerships with Tier1/2. Otherwise, the topic of release in our use case without these partnerships would be difficult to perform.

Further partnership with Hessen Mobil and the federal highway research institute (BASt) as state authority, simplify the approval process greatly.

CAN over Automotive Ethernet for Trailer Interface

Andreas Goers(✉) and Sebastian Kühne

WABCO GmbH, Mannheim, Germany
andreas.goers@wabco-auto.com

1 Introduction

Automated driving with commercial vehicles (CV) faces additional challenges due to a possibly attached trailer. In Europe approximately 65% of the CVs are towing trailers.

There are blind spots around the truck trailer combination which are not observable even when driving straight ahead. If an environment supervision is necessary to detect objects behind and next to the trailer, appropriate sensors must be installed on the trailer. Finally the sensor data must be transmitted to the truck.

Trucks and trailers are produced by different manufacturers and often unpredictably matched in service, so that a standardized interface is highly recommended.

For the data transmission between truck and trailer both wired and wireless transmission lines are possible in principle. Both have advantages and disadvantages. There will be use cases were wireless is better suited than wired transmission and vice versa (Figs. 1 and 2).

Fig. 1. Possible transmission paths

T. Bertram (Hrsg.): *Fahrerassistenzsysteme 2018,* Proceedings, S. 166–177, 2019.
https://doi.org/10.1007/978-3-658-23751-6_16

	Ethernet link Wired	WLAN IEEE 802.11
Bandwidth	high	medium
Latency	low	low
Link Reliability	high	low
Security	high	medium
Time to market	standardization required	easy retrofit
Recommendation	for safety critical data and low latency	for FMS and big data integration

Fig. 2. Comparison between wired and wireless

In the following the question about a wired transmission, with special attention on the existing vehicle interfaces, is investigated. In particular standardization is important to allow OEM independent usage.

2 State of the Art Truck Trailer Interfaces

In Europe two connectors between truck and trailer are mandatory.

1. ISO-7638 connector for ABS and EBS brake control on heavy duty trailers[1].
2. ISO-12098 connector for controlling the trailer lights on heavy duty commercial trailers[1] (Figs. 3 and 4).

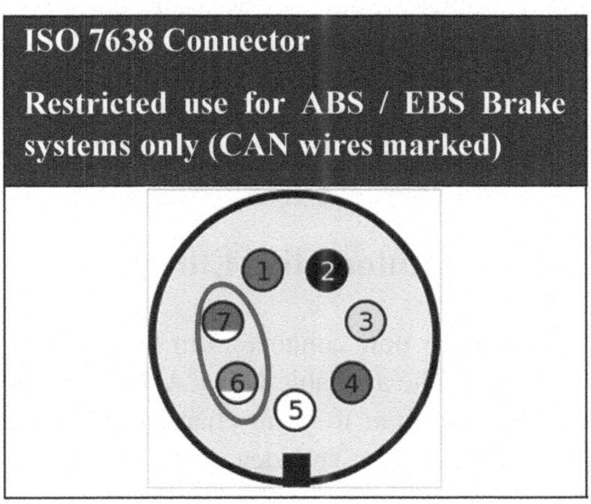

Fig. 3. ISO-7638 connector

[1]Heavy duty trailers and commercial heavy duty trailers are within the vehicle catego-ries O3 and O4 according to ECE/TRANS/WP.29/78/Rev.3.

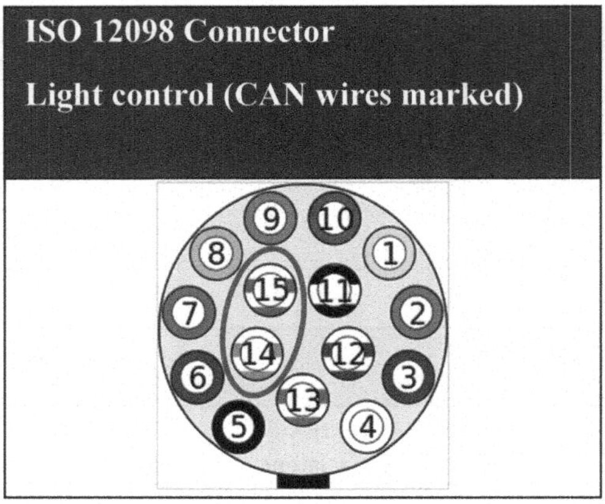

Fig. 4. ISO-12098 connector

For truck – trailer connections a low speed CAN channel according ISO-11992 is specified. The key features of a low speed CAN according to ISO-11992-1 are:

- Point to point connection (In difference CAN bus according to ISO 11898-2[2]).
- 1-wire capability (Full speed transmission even if one wire is open or shorted to GND or Ubat)
- 12/24 V voltage physical layer with maximum 125 kBaud transmission rate

The usage of the ISO-CAN busses is strictly regulated. For instance the data transmission on the ISO-CAN of the ISO 7638 connector is restricted to ISO-11992-2 defined messages which concern the braking system only. In addition the legal regulation ECE-R13 refer to the messages of ISO 11992-2.

ISO 11992-3 defines all data content which is allowed to be transferred via the ISO 12098 connector. Unfortunately, not all trucks and trailers support the ISO-CAN on the ISO 12098 connector yet. There is a new release of the ISO-11992-3 under review, where messages for the "Automatically Commanded Steering Function" (ACSF) are added. With this new release the speed and distance of a maximum of eight objects behind and beside the trailer will be transmitted.

3 Concept of CAN over Automotive Ethernet

In general it is hard to introduce new connectors to the Commercial Vehicle market because of the associated cost. The available ISO CAN is the perfect match for Automotive Ethernet because of the point to point characteristic (P2P) and the available unshielded twisted pair (UTP) harness. The idea is:

[2]ISO 11898-2 uses a two wire transmission path with up to 1 Mbit/s and more than two nodes. It is the most used physical layer in vehicle powertrain applications and indus-trial control networks.

- Using the available harness for a much faster Automotive Ethernet transmission.
- Transmit all ISO-CAN frames as unchanged within TCP/IP frames over a CAN – Ethernet – CAN Bridge (Fig. 5).

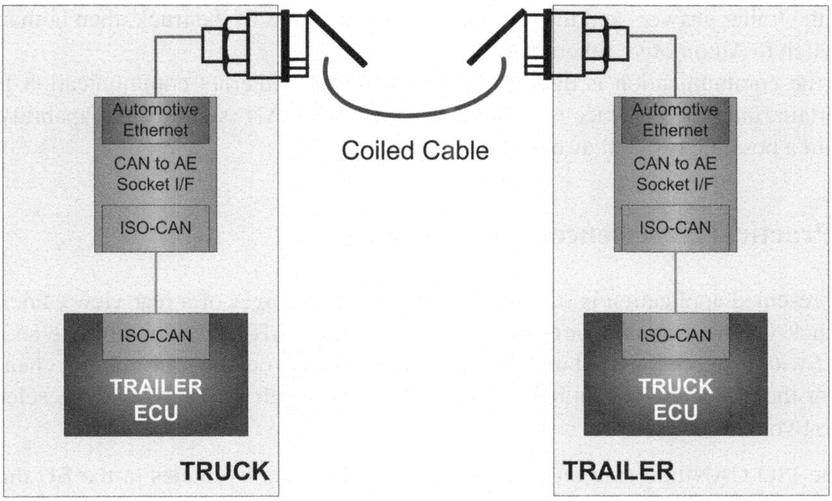

Fig. 5. CAN to Ethernet bridge

- Guarantee a low latency for bridged CAN messages with defined "Quality of Services" (QoS) according to IEEE 802.1.
- Fall back strategy to proven in-use ISO-CAN Interface.
- Gain bandwidth for additional data.

The concept presented here proposes to transmit all data given "as is" packed in Ethernet frames (42 Bytes … 1500 Bytes) and be sent via an Ethernet switch over the available wired transmission channel between truck and trailer.

For time critical data, mechanisms are standardized in IEEE 802.1, like "Time Sensitive Networking" (TSN) and "Audio Video Bridging" (AVB), to guarantee low reaction times in the range of 1 ms (Fig. 6).

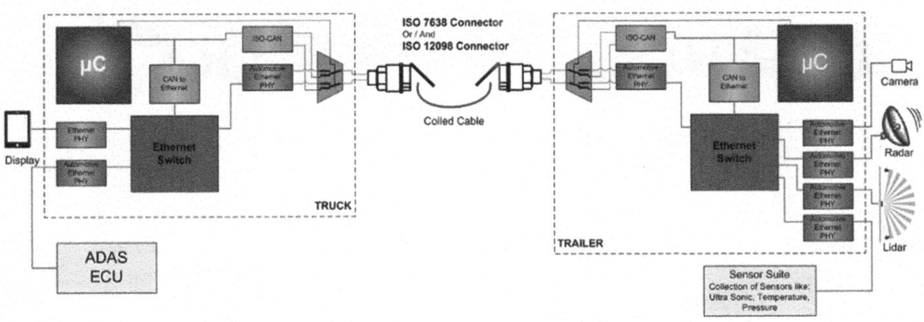

Fig. 6. Idea to gain additional bandwidth for environment sensors.

Concept how to negotiate the channel capabilities:

1. Both sides are starting with ISO-CAN communication as currently standardized. Negotiation can start only with an established two wire communication.
2. The truck queries the channel capabilities of the trailer with a diagnostic message. If the trailer answers and the capabilities fit to the ones of the truck, then both sides switch to Automotive Ethernet.
3. If the communication is disturbed (no Automotive Ethernet communication for a certain time) the interface is switched back to ISO-CAN (single wire capability) to gain a possible residual availability.

4 Practical Experiences – Test Results

The presented application is supposed to transmit the images of a rear view camera to the truck via the available harness. In addition to the CAN to Ethernet bridge an Ethernet Switch is necessary to bundle several data channels on one transmission channel.

For the in-vehicle feasibility test the transmission path via ISO 7638 was selected due to two reasons:

1. The ISO-CAN interface on ISO 7638 is available on all vehicles in the EU due to legal regulation whereas the availability of the ISO-CAN interface on ISO 12098 is not given.
2. The already existing data communication channel between the truck and trailer-EBS ECUs with repetition times down to 10 ms allows for the evaluation of the transmission channel with shared services.

The concept's feasibility was verified with a manual switch-over from ISO-CAN media to Automotive Ethernet on both sides (truck and trailer). First practical prototype tests were done with "unmanaged" Desktop Ethernet switches using RJ45 connector plugs. For transmitting the data via the vehicle harness between truck and trailer, media converters are used to test the transmission with Automotive Ethernet Physical Interfaces (PHYs) (Fig. 7).

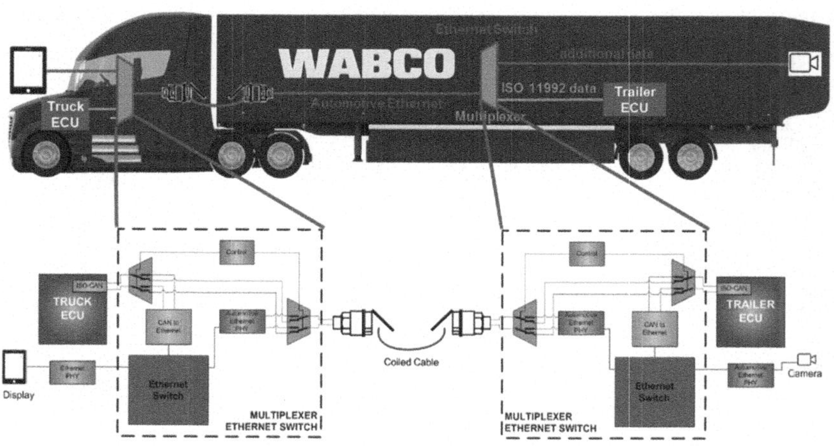

Fig. 7. Vehicle test setup (A-Prototype)

4.1 Harness Capabilities

The first attempt was a reuse of the available wiring harnesses and connectors. According to ISO 11992-1 the characteristic impedance is unspecified. Therefore, a new as well as an aged wiring harness, was tested with a network analyser. The wired transmission path is characterised with the line impedance Z_{RF} and the S-parameters. For the differential transmission performance the Insertion Loss (IL) and the Return loss (RL) is important. The IL is a measure for the attenuation of the signal when traveling from the transmitter to the receiver. The RL is a measure of the echo strength of the own transmission (Figs. 8 and 9).

	Lower Limit	Upper Limit
CIDM Z_{RF} [Ω]	90	110
IL S_{dd21} @ 1 MHz [dB]	0,7	1,0
IL S_{dd21} @ 10 MHz [dB]	1,8	2,6
IL S_{dd21} @ 33 MHz [dB]	3,4	4,9
IL S_{dd21} @ 66 MHz [dB]	5,0	7,2
RL S_{dd11} @ 1 MHz [dB]	18,0	23,0
RL S_{dd11} @ 2 MHz [dB]	18,0	23,0
RL S_{dd11} @ 2 MHz [dB]	18,0	30,0
RL S_{dd11} @ 20 MHz [dB]	18,0	30,0
RL S_{dd11} @ 66 MHz [dB]	12,8	24,8

Fig. 8. IEEE 801.3bu BroadR-Reach definition for the communication channel

The red line in Fig. 10 represents the upper limit of the insertion loss as specified by the Open Alliance BroadR-Reach (OABR).

Result of the first cable analysis: Regarding the Inversion Loss ($IL = S_{21}$) the wiring harness is feasible as Automotive Ethernet transmission path. Additional investigations are necessary to make a general statement.

4.2 Latency Measurements

Two different latencies were measured

1. "Glass to Glass" latency which means: Camera lens ("Glass") → transmission path → Laptop as Display ("Glass"). The tested Automotive Ethernet camera alone is specified with worst case 55 ms latency. With some adjustments and optimizations in the video playback software "gstreamer" a typical summarized latency of 100 ms was achieved.

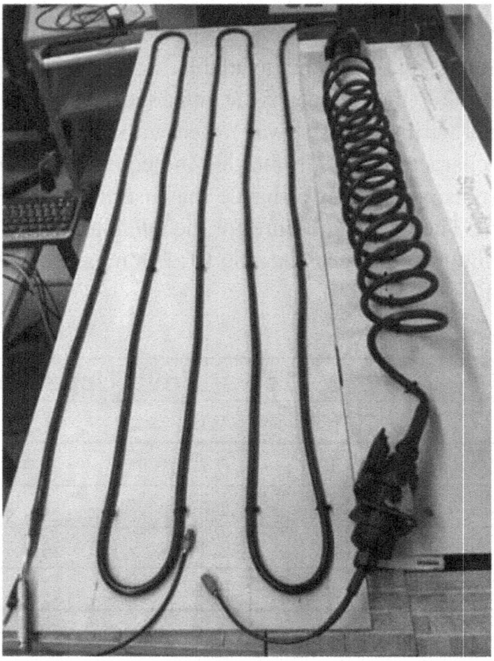

Fig. 9. Device under test: The cable

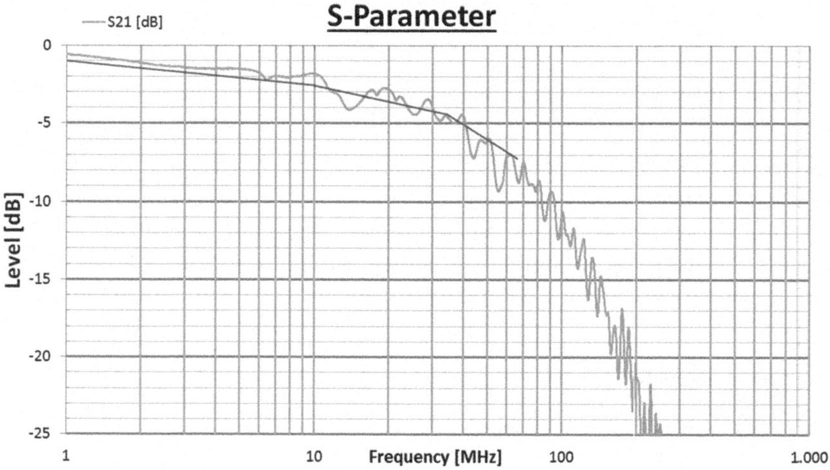

Fig. 10. Insertion Loss = S21 parameter. (Color figure online)

2. CAN latency due to conversion, switch latencies and reconversion. With this test the latency of the transmission path can be assessed. Test setup:
   - CANalyzer sends out CAN Messages on one end of the transmission path and receives the CAN Messages on the other end of the transmission path. The time delay is the "CAN latency" measured as shown in Fig. 11 below.
   - The "CAN latency" measured with two additional cameras streaming their pictures over the transmission path.
   - Finally, a network stress test is added with a network throughput test program "iPerf3". All remaining network capacity is occupied by this stress test, representing the worst case for the "CAN latency" measurement (Fig. 12).

Setup	CAN CycleTime [ms]	CAN Busload [%]	CAN Latency [ms]	Lost CAN Messages? Errorframes?	iPerf3 Data transfer [Mbit/s]	Camera 1 Video On?	Camera 2 Video On?
CAN Only	10	12,7	1,85	NO	NO	NO	NO
CAN_Multiple_Messages(8)	3x10 / 4x100 / 1x1000	40	2	NO	NO	YES	NO
CAN_Multiple_Messages(8) and ABUS andiPerf3	3x10 / 4x100 / 1x1000	40	2,3 - 7	NO	93 - 94	YES	NO
CAN_Multiple_Messages(8) and 2xCameras	3x10 / 4x100 / 1x1000	40	2,3	NO	NO	YES	YES
CAN_Multiple_Messages(8) and 2xCameras +iPerf3	3x10 / 4x100 / 1x1000	40	1,7...5,5..8,1	NO	86-91	YES	YES

Fig. 11. CAN latency measurements test results

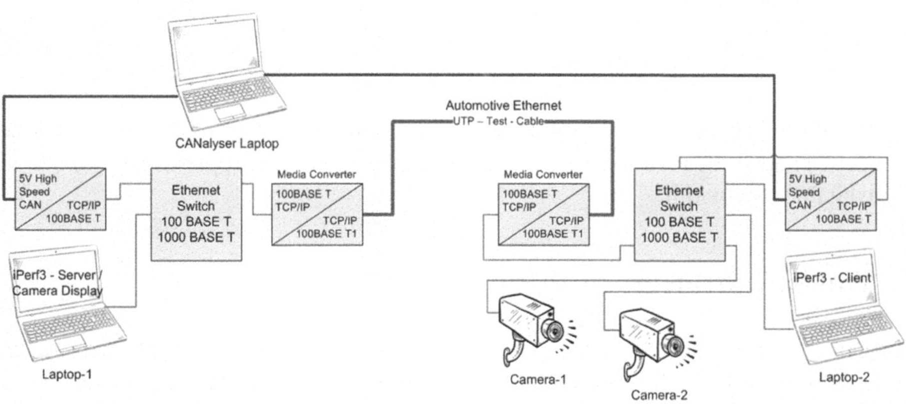

Fig. 12. Test setup for CAN latency measurement on test bench

The measurements revealed that the Ethernet routing of the CAN messages must be optimized with Time-Sensitive Networking (TSN) to gain a constant and low latency. To guarantee a transmission of safety critical data within a specific time different IEEE standards for different use cases are added within the last years.

4.3 In-vehicle Tests

The described A-Prototype setup was tested under "real-life" conditions on the WABCO test track. Without any changes on the EBS ECUs on truck and trailer it was possible to transmit all ISO-11992-2 data over the Automotive Ethernet bridge while receiving the stream of the rear view camera. The original wiring (trailer power cable and coiled cable) was used (Fig. 13).

Fig. 13. Trailer box with A-Prototype setup

The ISO-CAN communication between truck and trailer was free of errors and the occupied bandwidth for the rear view camera was measured approximately 30% of the available bandwidth.

5 Fail Safe Strategy

Currently the Low Speed CAN is able to cope with failures on a single wire (open load, short, short to Gnd, short to Ub) because of the One-Wire operation mode

The general fail-safe strategy with the shown "CAN over Automotive Ethernet" approach is: In case of any transmission path failure, switch back to ISO-CAN (low-speed CAN) to utilize the more robust CAN failure handling (e.g. single wire failures) and retain a working transmission channel for the legal recommended functions. As defined in ISO-11992-1 the fault handling procedure starts when data transmission was not possible for a certain time. Similar state machines with different timings are mandatory for the towed and the towing vehicle, to ensure that the communication is established again after a certain time.

This proven method can be extended with the "CAN over Automotive Ethernet" proposal: If the Automotive Ethernet communication is not possible or below a defined quality level, then the transmission path is switched back to ISO-CAN starting the fault handling procedure to ensure the transmission of safety relevant data at least.

In parallel to the supervision of the data flow, the voltage level of the wiring can be monitored. In case the SW timeout supervision turns out not to be sufficient for the considered safety case, bus failure detection as well as CAN transmission levels can be supervised with an independent monitoring path.

6 Approach for ISO Extension

As mentioned before, the truck trailer communication is regulated by ISO-11992. The ISO Standard includes four parts

- Part 1: Physical layer and data-link layer
- Part 2: Application layer for brakes and running gear
- Part 3: Application layer for equipment other than brakes and running gear
- Part 4: Diagnostics

WABCO prepares a change request with the idea to extend the ISO standard with an additional Part 5: *Additional Physical layer and data link layer*. With this extension two main market demands will be satisfied:

1. Backward compatibility: If one of the communication partners is only able to communicate with ISO-CAN, then no additional communication channels are possible but all legal requirement are still fulfilled.
2. Introduce a high-speed communication channel and open the wired interface for other participants. Define a maximum latency for transmission of the ISO content. Define the fail-safe strategy.

Of course, the planned extension will be available for the ISO 7638 connector as well as for the ISO 12098 connector

7 Conclusion and Outlook

The feasibility of the concept "CAN over Automotive Ethernet for the trailer interface" is given. The successful in vehicle test in conjunction with the promising evaluation of the transmission path rates the shown concept as feasible. WABCO will go ahead to investigate the transmission path in more detail.

7.1 Short Term Outlook

The next step in the evaluation is the extension of the concept to up to five trailers using the CAN communication path via ISO 7638 connector. The planned prototype is using an Automotive Ethernet switch with Audio Video Bridging (AVB) and Time-Sensitive-Networking (TSN) capabilities. As partner in the EU-Project

AEROFLEX (https://aeroflex-project.eu/) WABCO supports the truck trailer communication using an Automotive Ethernet capable Router Repeater Unit.

To ensure a low latency the CAN to Ethernet bridge function is integrated in the Router/Repeater Microcontroller. In addition a high speed CAN interface is foreseen to support trailer related control functions for the AEROFLEX project (Fig. 14).

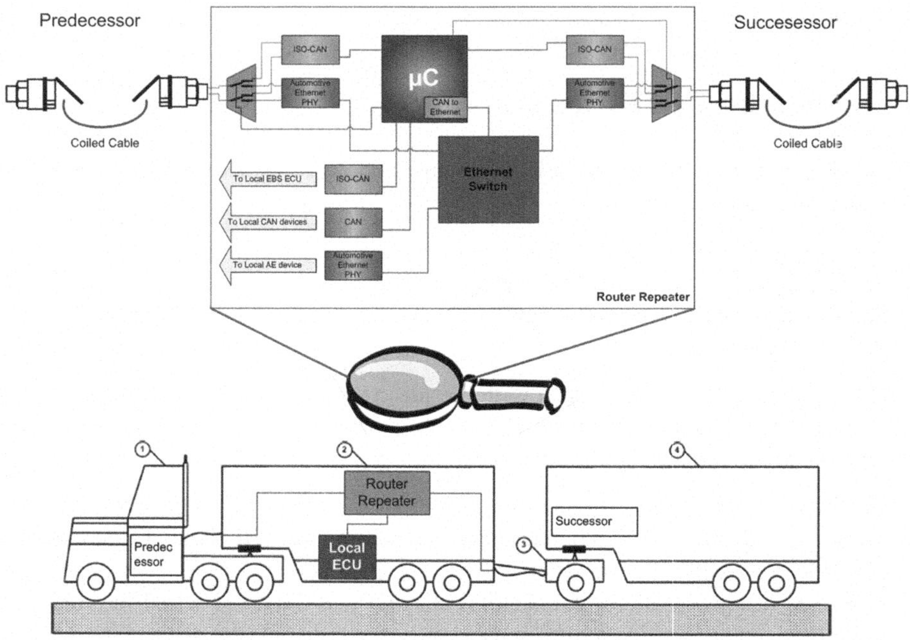

Fig. 14. Next step of prototype: Automotive Ethernet Router Repeater for up to 5 trailers

7.2 Long Term Outlook

In the long term a cost-effective solution for integration in the available truck and trailer ECU's is targeted (Fig. 15).

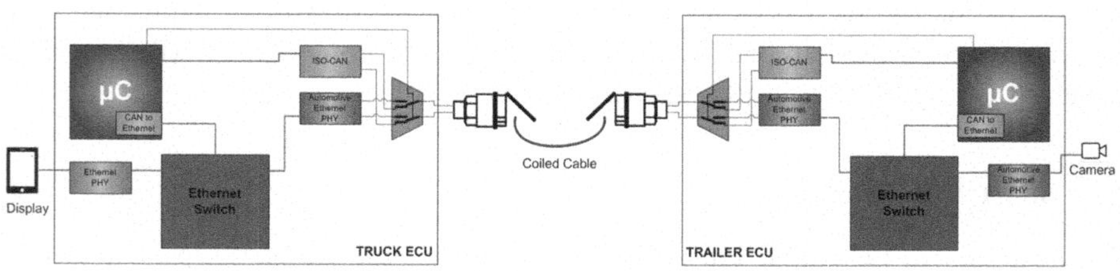

Fig. 15. Integration in truck and trailer ECU's

The number of additional components in the trailer using Automotive Ethernet like Cameras, Radars, Lidars and so on, may vary depending on the necessary function. One big advantage of Ethernet is, that the extension of the network is possible using a separate Automotive Ethernet switch device where additional devices can be attached.

8 Summary

Future truck automation will require high bandwidth data communication with maximum reliability and robustness which is only achievable with wired communication. The presented proposal will allow to reach these targets without adding connecting cables between truck and trailer. The broadest benefit will be achieved when this not only stays a proprietary solution, but becomes a future standard.

An Overview of Deep Learning and Its Applications

Michael Vogt(✉)

Smiths Heimann GmbH, Wiesbaden, Germany
michael.vogt@smithsdetection.com

1 Introduction

Deep learning is the machine learning method that changed the field of artificial intelligence in the last five years. In the view of industrial research, this technology is *disruptive*: It considerably pushes the border of tasks that can be automated, changes the way applications are developed, and is available to virtually everyone.

Subject of deep learning are artificial neural networks with a large number of layers, which motivates the name. Compared to the earlier approaches with (ideally) a single layer, this allows training black-box models directly on raw data with a minimum of engineering work, called *end-to-end* learning. Most successful applications are found in image, audio and text modeling. This report will find out what made this breakthrough possible and what can be expected from it.

2 Artificial Neural Networks

The beginning of neural network research dates back to the middle of the 20th century. In 1943, McCulloch and Pitts [57] published a first model for biological neurons shown in Fig. 1: Input signals xi are weighted and summed up in the cell body, followed by a nonlinear activation function φ determining the cell when to "fire". Variants of this model are still the processing units of today's deep networks.

The next bigger milestone was Rosenblatt's perceptron [71] in 1958, which is a network of neurons with connections directed from an input to an output layer, see Fig. 2. This configuration distinguishes *feed-forward* networks from other types like recurrent networks (Sect. 8) or self-organizing maps. In the case of multiple hidden layers it is called a multi-layer perceptron (MLP).

Feed-forward networks represent functional mappings between input and output. *Training* (or learning) means determining the weights of all layers based on empirical data; if the desired outputs are known, it is called *supervised* learning. The error back-propagation algorithm (EBP) [72, 89] does this by minimizing the squared error on a training data set using gradient descent. Each step updates the weights in negative gradient direction with the learning rate η, summarized for a single neuron in Fig. 3.

T. Bertram (Hrsg.): *Fahrerassistenzsysteme 2018*, Proceedings, S. 178–202, 2019.
https://doi.org/10.1007/978-3-658-23751-6_17

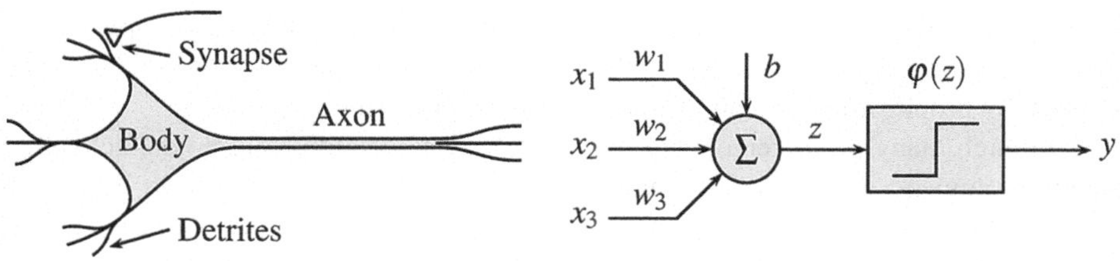

Fig. 1. Simplified biological neuron (left) and McCulloch-Pitts model (right)

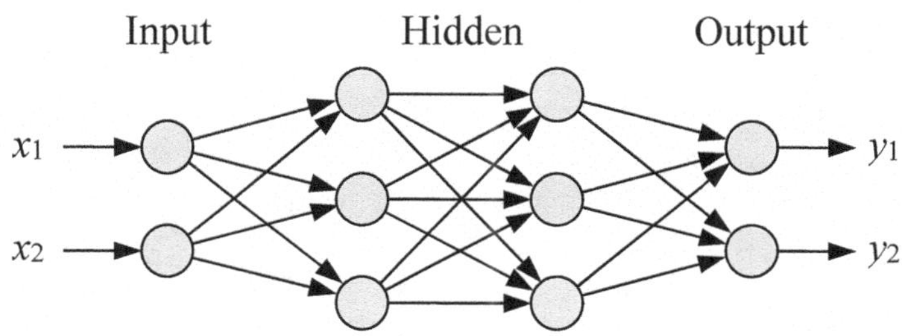

Fig. 2. Feed-forward neural network with two hidden layers

Neuron: $$y = f(\mathbf{x}) = \varphi\Big(\sum_i w_i x_i + b\Big) = \varphi(\mathbf{w}^{\mathrm{T}}\mathbf{x} + b) \tag{1}$$

Data: $$(\mathbf{x}_k, t_k) \quad \text{for } k = 1, \ldots, N \tag{2}$$

Loss: $$V = \sum_k V_k = \frac{1}{2}\sum_k e_k^2 = \frac{1}{2}\sum_k (f(\mathbf{x}_k) - t_k)^2 \tag{3}$$

Gradient: $$\frac{\partial V}{\partial w_j} = \sum_k \frac{\partial V}{\partial e_k} \cdot \frac{\partial e_k}{\partial z_k} \cdot \frac{\partial z_k}{\partial w_j} = \sum_k e_k \cdot \varphi'(z_k) \cdot x_{jk} \tag{4}$$

Update: $$w_j^{\text{new}} = w_j^{\text{old}} - \eta \cdot \frac{\partial V}{\partial w_j} \tag{5}$$

Fig. 3. Gradient computation and weight update for the single neuron from Fig. 1

The mapping of a full layer with p neurons is described as

$$\mathbf{y} = \begin{pmatrix} y_1 \\ \vdots \\ y_p \end{pmatrix} = \varphi \begin{pmatrix} \mathbf{w}_1^{\mathrm{T}}\mathbf{x} + b_1 \\ \vdots \\ \mathbf{w}_{\mathrm{p}}^{\mathrm{T}}\mathrm{x} + b_p \end{pmatrix} = \varphi(\mathbf{W}\mathbf{x} + \mathbf{b}) \tag{6}$$

where $\mathbf{y}$ is the layer's output vector, i.e., the outputs of the p neurons. Applying the chain rule repeatedly, EBP computes the derivatives with respect to all weights in the network by propagating the output error back from layer to layer.

Although many improvements have been added, like smooth sigmoidal activation functions (logistic or hyperbolic tangent, see Fig. 8) and momentum term [68], with these tools it is nearly impossible to train networks having more than one layer. As it was shown at the same time that networks with a single layer are *universal approximators* [14, 35], research in the 1990s was limited to these "shallow" networks.

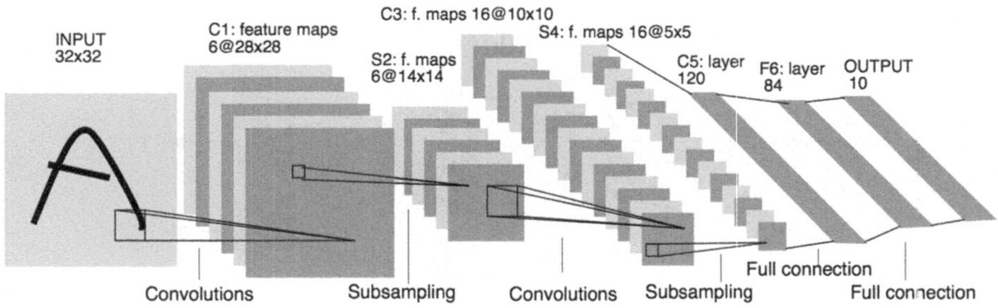

Fig. 4. The LeNet-5 architecture; original drawing from [47]

However, there was a successful exception: "LeNet" [46, 47] in Fig. 4 was designed for pattern recognition in images and has been used for reading handwritten numbers on checks in the US over many years. It is composed of two standard layers preceded by so-called *convolutional* layers that extract features by applying small filters to the input data. LeNet was the first deep convolutional neural network.

3 Convolutional Neural Networks (CNNs)

After the "AI winter" during the 2000s, neural networks had an unexpected revival: In 2012, the achieved error rate in the yearly "ImageNet large scale visual recognition challenge" (ILSVRC [73]) dropped from 26 to 16% – and the winner [43] was a convolutional network similar to LeNet.

3.1 Components of Convolutional Networks

Deep convolutional networks (CNNs) are characterized by a sequence of convolutional and pooling layers, followed by one or two fully connected layers, as in Fig. 4. A convolutional layer is a particular type of layer for image and signal processing, different from the perceptron-style layers in Sect. 2. It is inspired by Nobel Prize winning studies [37] showing that neurons in the visual cortex have only small receptive fields.

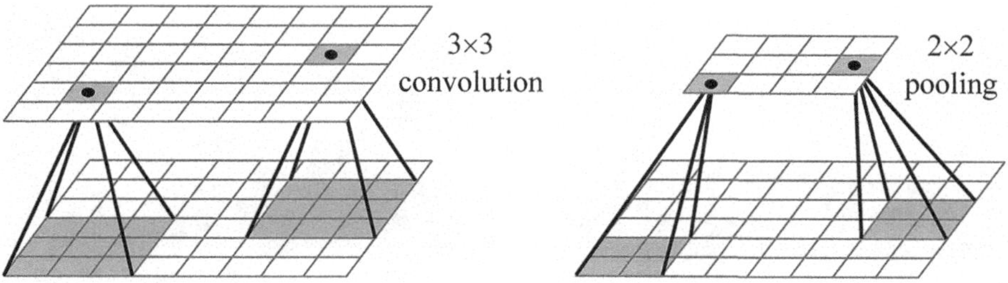

Fig. 5. Basic CNN layers – convolution (left) and pooling (right)

This is reflected by connecting each neuron only to a small neighborhood of image pixels and sharing the weights among all neurons of a particular *feature map* (Fig. 5, left panel), which is equivalent to convoluting the image with a small filter kernel. It shrinks the number of weights to the filter size and makes them independent of the position in the image. Note that each convolutional layer learns multiple of these feature maps, indicated in Fig. 4. The layer inputs and outputs are commonly modeled as multi-plane 2D data rather than a long vector in the general case of Fig. 2.

The second construction element of CNNs are *pooling* layers, which perform subsampling of the feature maps by simple operations like maximum or mean. The importance of pooling is twofold: First, it reduces the input size for the next layer (Fig. 5, right panel) and allows learning more mappings instead. And second, it helps combining the output from the previous layer on a coarser scale.

The deep part of the CNN can be seen as a feature extractor: The first layer detects, e.g., edges and corners at pixel level; the next layers indentify textures, object parts, and so on. By construction, the spatial relationship is propagated through all convolutional layers; the fully connected layers finally learn the associated classes and predict probabilities for the object contained in the image. All weights of this network are trained simultaneously. While hand-crafted features used to be the clue to the art of object recognition over decades, they are now learned from the raw image data.

3.2 Image Classification with CNNs

There is a variety of publically available data collections, often provided for yearly challenges, where teams can submit solutions to be evaluated on independent test data. The most noted one in the last years is the above mentioned ILSVRC. One task is to classify objects in images, based on a training set of 1.4 million instances in 1000 classes. The performance is assessed by the top-5 error rate, i.e., an image is correctly classified if the correct class is among the 5 most likely proposals, see Fig. 6.

Fig. 6. ILSCRV example images with top-5 proposals; original results from [43]

Table 1. Winner algorithms of ILSVRC 2010–2017

Year	Error (%)	Depth	Algorithm
2010	28.2	Shallow	NEC Labs
2011	25.8	Shallow	Xerox Research
2012	16.4	8	AlexNet [43]
2013	11.7	8	ZF Net [95]
2014	6.66	22	GoogLeNet [84]
2015	3.57	152	ResNet [30]
2016	2.99	200	Ensemble of CNNs
2017	2.25	152	Squeeze-and-Extract

Table 1 summarizes the results since 2010, the era of CNNs starting with "AlexNet" [43] in 2012. The subsequent years yielded two important advancements that are state-of-the-art today:

- Inception modules, which are the building blocks of GooLeNet [84] and the Inception series [85, 86]. They can be considered as complex layers able to extract features at different scales.
- Residual connections that allow the ResNet structure [30] building very deep networks by propagating inputs over multiple layers.

Both mechanisms aim at making the networks deeper while keeping them trainable; they have recently been joined to Inception-ResNet [86]. Another popular architecture is the 2014 runner-up VGG [80] with its simplistic composition of layers (Fig. 7).

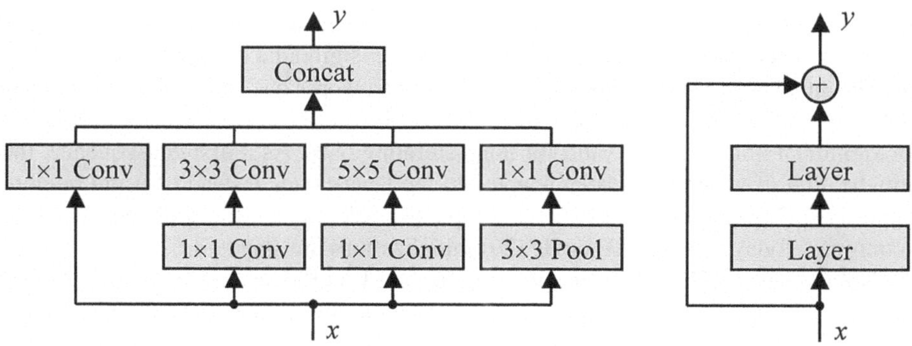

Fig. 7. Inception module v1 (left) and residual module (right)

Within only a few years the error rate fell from 26% to less than 3%, which is beyond the human performance of 5%. The importance of ILSVRC is that it helped identifying the best-performing approaches, most of which are made available on platforms like Github. Nearly all relevant applications today use these networks or modifications; they are the basis for more complex tasks like object detection in Sect. 7.

4 Training Deep Neural Networks

Section 3 described the concept of CNNs and their triumph in image classification. As the major components were already present in LeNet in the 1990s, the question remains: What changed in the following 20 years leading to this success? The next sections will identify the main ingredients, starting with algorithms.

4.1 Vanishing Gradients

A major obstacle in training deep networks is the "vanishing gradients problem". As the EBP algorithm goes to deeper layers, the originally used sigmoidal activation functions tend to saturate, such that the gradient shrinks to zero. With the machinery of Sect. 2, it is virtually impossible to train those networks. One way of mitigation is using non-saturating functions.

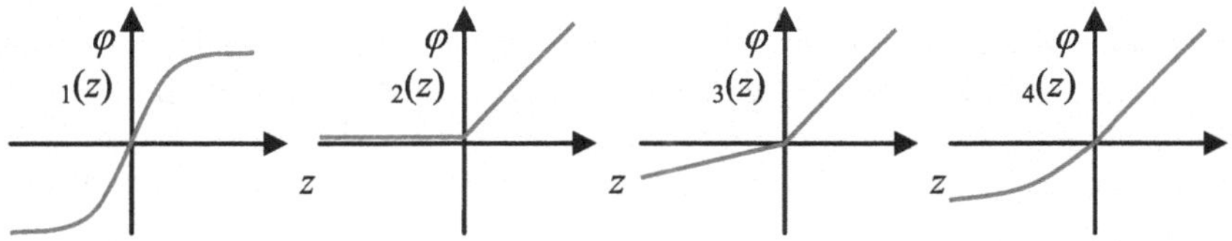

Fig. 8. Activation functions – (1) hyperbolic tangent, (2) ReLU, (3) leaky ReLU, (4) ELU

The first proposal [24] was the *Rectified Linear Unit* (ReLU), see Fig. 8(2). Despite its non-smoothness, it accelerates the training and enables the weights in deeper layers to be adapted. Improved variants have been introduced like the Leaky ReLU [93]

or the ELU [11], all with their pros and cons. Independent of the specific type, non-saturating functions are nearly always preferred over sigmoidal ones in CNNs.

The analysis of the data flow in feed-forward networks has shown that it is favorable keeping the inputs and outputs of each layer in similar ranges. This is mandatory for sigmoidal units but also valid for non-saturating ones. As a first consequence, the initialization of weights has been adapted to the layers sizes instead of using random values [23]. Together with ReLUs, this simple trick already led to significant speed-up in training. Today, Xavier [23] or He [29] initializations are standard.

Batch normalization [39] goes one step further: It introduces a scaling layer before the activation functions and learns the optimal scaling from the data batch in use. This again improves both accuracy and training speed by tackling the vanishing gradients. According to [39] it is then even possible to train deep networks with sigmoidal units.

4.2 Optimization Methods

Regarding the optimization algorithms, there is also some progress since the times of EBP. As a main difference, the gradient is not computed on the full data set but on stochastically chosen "mini-batches". Thus, all methods can be regarded as forms of *stochastic* gradient descent (SGD). For the huge networks in Sect. 3, this has mainly practical reasons: Training is usually performed on a GPU (Sect. 6.1), and for computing the gradient all intermediate values of the network must be stored. The mini-batch size is then naturally determined by the available GPU memory. Among the algorithms, there are two branches:

- Non-adaptive methods: Original EBP [72], momentum optimization [68], Nesterov's accelerated gradient [60]
- Adaptive methods: AdaGrad [19], RMSProp, Adam [42]

Finding the optimal values for learning rate, momentum etc. is an open issue. It is common to repeat the training with a variety of settings. Adaptive methods scale the gradient component-wise based on the data variance and thereby promise more robust parameter selection. However, it has been found that this class may lead to worse generalizing networks [92], so that the competition is still open. Currently, the preferred choices are either Nesterov's method (non-adaptive) or Adam (adaptive).

4.3 Regularization

The networks in Sect. 3 typically have millions of parameters. It is beyond discussion that there will never be enough data available to really parameterize them, not even rudimentarily. Prohibiting overfitted solutions is only possible with a strong and workable *regularization*. Besides well-known devices like "early stopping", there are mainly two advancements:

- CNNs by construction. The strong coupling of weights in the convolutional layers makes their number dependent only on the sizes of the filter kernels. Some CNNs use this as sole regularization method.

- *Dropout* [81] is as simple as efficient: In each pass of the training, a randomly chosen fraction of neurons is completely ignored. This enforces solutions that are robust against changes in the data. Dropout rates of 50% are not uncommon.

Summing up this section, a lot of single improvements have been made to the original EBP, allowing now the training of deep networks, which was not possible before.

5 Dealing with Data

Regardless of the tools described in Sect. 4, training all layers of a deep network requires substantially more data than other machine learning methods. As pointed out earlier, the ILSVRC provides more than one million images to the participants. Usually, there is much less training data than desired, so the goal must be to use it as efficient as possible.

5.1 Importance of Labeled Data

The availability of large labeled data sets (Table 2) is central for the advancements in deep learning. Such collections are prepared in funded projects or by research groups and made available to the public to be used for algorithm development and assessment. The decisive aspect is that this data is *labeled* (or *annotated*), i.e., it contains class labels, object positions, translations, or whatever algorithms should predict. For visual tasks, annotations range from plain class descriptions and bounding boxes to complete semantic labeling. Data sets like "Citiscapes" [12] are necessary for scene understanding in autonomous driving and alike, see Fig. 9.

It is obvious that gathering labeled data in the field is of particular value: Photo databases containing keywords, faces associated with names on social media platforms, user ratings of products or movies. If annotations are not present, companies invest amounts of time and money in, e.g., marking objects in images or video sequences. The ImageNet data provides a good example how such a process is set up with "crowd sourcing", ensuring label quality and other devices [73, 82]. For acceleration, preliminary detection algorithms can aid the labelers in a way that only missing or wrongly assigned labels need to be corrected. Today, there are many platforms [13] and even commercial companies [1] offering data annotation services.

Table 2. Popular labeled data collections for visual tasks

Data set	Instances	Description
ImageNet	1.4 m	Images of objects, 1000 classes, used in ILSVRC [73]
CIFAR	60 k	Images of objects, 10 or 100 classes, often for testing
Pacsal-VOC	500 k	Images of objects, includes bounding boxes
COCO	2.5 m	Images of complex scenes (Microsoft)
MNIST	60 k	Handwritten digits, available since 1998, see [47]
Cityscapes	5 k+20 k	Semantically labeled urban scenes [12]
GTSRB	50 k	German traffic signs, 40 classes
LFW	13 k	"Labeled faces in the wild", 1680 multiple occurrence

Fig. 9. Semantic labeling of urban scenes in the "Cityscapes" database [12]

This constitutes an important change in the development process for new products: The efforts are shifted from designing algorithms to managing data. As described in Sect. 3, many of the best-performing networks including their training software are released to the public, and usually it is not promising to compete against these networks with own developments. It is much more important having people for gathering, labeling, and preparing sufficient training data. This insight is not new [4] but gained much more practical relevance with the appearance of deep learning.

5.2 Using Generated and Unlabeled Data

As a basic principle, nothing is more valuable for current training algorithms than labeled data. However, there are several ways for bringing down the amount needed. One direction is artificial *augmentation*; in the case of images, this means:

1. Cropping, flipping, scaling, see Fig. 10
2. Change lighting, contrast, etc.
3. Inject (overlay) objects to be found into images

The first group of operations is often included in training frameworks for easily augmenting the data by a factor of 10 or more. The other two operations need to be carefully implemented to prevent the network from learning the differences between

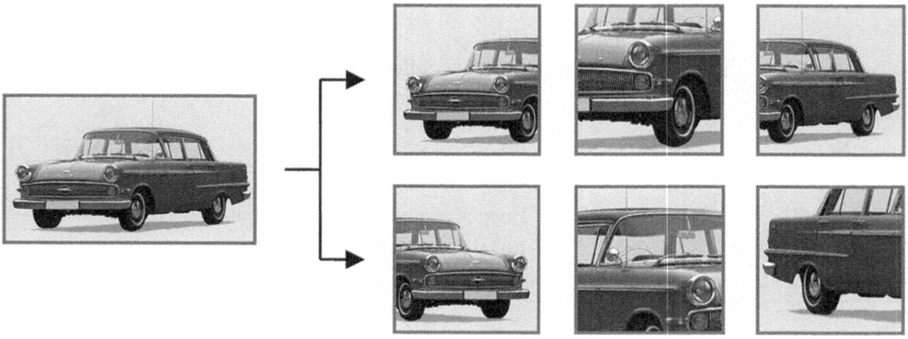

Fig. 10. Image augmentation by geometric operations (image from https://pixabay.com/)

natural and artificial images. All these efforts have in common that they intend to get more data without manual labeling.

Another approach is exploiting *unlabeled* data. This comes in two flavors:

- Unsupervised pre-training. Modeling the data in each layer as autoencoder or "restricted Boltzmann machine" was one of the first approaches for training deep networks [32]. An alternative is end-to-end training of a *deep* autoencoder [96].
- Self-supervised training on an auxiliary (or proxy) task, for which labels can be generated automatically. The underlying notion is that the deep layers extract similar features independent of the classification task. Possible setups are:
  - Learning the spatial relationship of image tiles [17].
  - Inpainting, i.e., predicting the contents of image regions [65].
  - Colorization of grayscale images [44].

Unsupervised pre-training went a bit out of fashion, because of the advancements described in Sect. 4. Today, supervised training is preferred as long as enough labeled data is affordable.

A new branch are *generative adversarial networks* (GANs) [25] which aim for creating synthetic content with the same characteristics as given input data. Although still under heavy research, they have been successfully applied to semi-supervised learning [75]. For several data sets from Table 2, competitive results have been achieved with only a small fraction of the images being labeled.

5.3 Reusing and Adapting Networks

Transfer learning is a technique particularly useful for quickly adapting a system to new requirements. The theory is very similar as for the auxiliary tasks mentioned above: When, e.g., introducing a new class in object classification, the deep layers do not change much as the features remain similar for the new task. In the simplest case, the convolutional part of a CNN is kept fixed, only the final layers are re-trained on a rather small number of images. For adding a new object it may then be enough to have 100 images instead of 100000.

This can be used for teaching an object detection system new traffic signs, for adding new people to face recognition or for adapting speech recognition to local dialects. In a more general sense, transfer learning means using a network pre-trained on a certain task (e.g., on the data sets of Table 2) as a starting point and adapt it for own applications. Benefit is always the reduction of data and training time.

6 Deep Learning Platforms

Besides algorithms and data, the third component that enabled deep learning is the availability of fast GPU hardware and software to easily use it for neural network training. This is often regarded as the most important element.

6.1 Deep Learning Hardware

As explained in Sects. 2 and 3, evaluating a CNN consists mainly of convolutions and large matrix multiplications (6). Both operations are perfectly suited for GPUs with their large number of computational cores and high memory bandwidth. For gradient computation during training, the mini-batch sizes are usually adjusted so that all data fits into the GPU's memory, which easily leads to factors of 10 to 100 in speed compared to CPU implementations. The market is dominated by NVIDIA hardware together with CUDA [61] and cuDNN [62] libraries.

- For network *training*, many vendors offer hardware with multiple GPUs, either cheap consumer cards or more specific processors like NVIDIA Tesla. The software frameworks described below can distribute training tasks to multiple GPUs.
- For *deploying* neural networks, embedded systems are available in addition to standard GPU cards. NVIDIA Jetson TX [63] is a credit-card sized module for general applications; NVIDIA Drive PX [64] in diverse configurations particularly serves the needs of vehicle automation.
- First processors have been developed specialized for operations needed in neural networks, like Google's Tensor Processing Unit [27] or Intel Nervana [38].

6.2 Software Frameworks

During the last years, a variety of software libraries have been developed for configuring, manipulating and training neural networks; Table 3 shows only a selection. They are often accessible on a high abstraction level through script languages like Python [22] or graphical user interfaces. This allows building deep learning applications without much detailed knowledge.

Table 3. Popular open source deep learning frameworks

Framework	API	Maintainer	Started
Caffe	Python, C++, Matlab	UC Berkeley	2013
CNTK	Python, C++	Microsoft, Inc.	2016
TensorFlow	Python, C++	Google, Inc.	2015
Theano	Python	Univ. of Montréal	2010
Torch	C++, Lua	R. Collobert et al.	2002

The frameworks support training on multiple GPUs and deploying trained models to PCs or embedded systems, thereby enabling quick development cycles. Many research results and successful networks (e.g., from Sect. 3.2) are publically released and can be immediately imported and adapted. This is a tremendous speed in spreading improvements, and it makes the technology available to a large community, not only a small group of experts.

6.3 Cloud Services

Table 4 lists a number of vendors who have started to include deep learning capabilities into their cloud computing products. The offers start with services for well-defined use-cases like object detection or text translation. They are usually accessed by HTTP/S requests implementing the REST paradigm. Data (images, text) is send to the service's URL, which returns the result in JSON or XML formats. Also training networks on own data is possible.

Table 4. Selection of commercial cloud services offering deep learning capabilities

Vendor	Product	Main entry point
Amazon	Web Services	https://aws.amazon.com/
Google	Cloud Platform	https://cloud.google.com/
IBM	Bluemix/Watson	https://www.ibm.com/watson/
Microsoft	Azure	https://azure.microsoft.com/

At the upper end, some services include web-based development environments, e.g., Microsoft's "Azure Machine Learning Studio" or Google's access to its TensorFlow framework (Table 4). The services are charged per computation time, storage, or number of evaluations. As main benefit, users can easily start without own infrastructure and scale it as needed.

7 Object Detection and Localization

The CNNs described in Sect. 3 perform in essence a *classification* task: They assign an object class to the complete image assuming that it contains only one primary object. Detecting objects in scenes is the next step in complexity for which the system needs to determine both position and class as in Fig. 11. Of course, this is much more valuable for practical applications.

A straightforward way would be using a segmentation method or a sliding window to define regions of interest (ROIs) and then feed them into a CNN. However, this is inefficient at run time as the CNN needs to be evaluated for each ROI. Instead, the ROI can simply be learned. As discussed in Sect. 5.2, features are usually similar for different tasks, thus ROI and class probability can even be predicted with a common network. There are two paths to do this: ROI proposals and single-shot detectors.

7.1 Predicting Object Locations

The *ROI proposal* approach is associated with a series of algorithms called R-CNN, Fast R-CNN and Faster R-CNN [70]. Based on a common stack of convolutional layers (consuming most of the computation time, see Fig. 12) two tasks are performed:

- Estimating the ROI coordinates in a region proposal network. The number of proposals is critical to the network's performance; common choices are 300 or more.
- Extracting the ROIs (= ROI pooling) and passing them to the final classifier.

Typical *single-shot* detectors are YOLO [69] and SSD [53]. Taking the feature maps from the convolutional layers, they simultaneously estimate the classes and offsets for a fixed number of bounding boxes. SSD improves this by additional convolutional layers helping to locate objects on different scales (Fig. 12, right panel). This produces a large number of confidence values for box/class pairs, from which the final decision is taken by non-maximum suppression.

Fig. 11. Localizing objects by frames (left, [70]) or masks (right, [31])

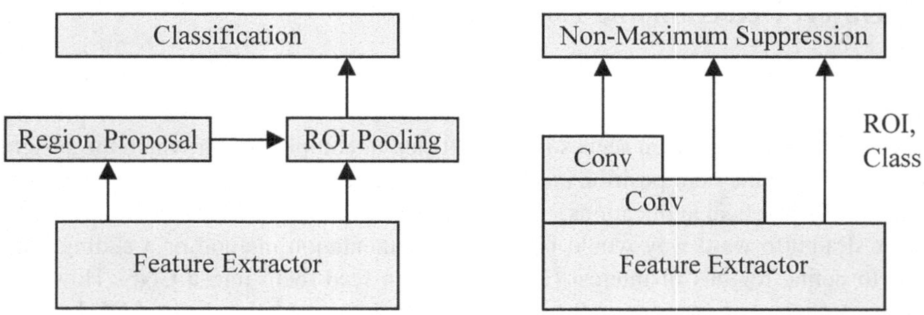

Fig. 12. Structures for object detection – ROI proposal (left) and single-shot (right)

Single-shot detectors are generally faster but less precise compared to ROI proposal methods, both using the CNNs from Sect. 3 as underlying structure. Google has released a common interface [36] that allows combining CNNs with one of the object detection meta-structures Faster R-CNN, SSD, or R-FCN [15]. From the systematic evaluation in Fig. 13, one can select the right balance between precision and speed fulfilling the requirements of a particular application.

Latest addition is the "Mask R-CNN" [31] which locates objects by masks instead of bounding boxes (Fig. 11) and is thus a form of semantic segmentation [51, 54].

This demonstrates how deep learning leverages image processing domains beyond classification – like segmentation, retouching [65], or up-scaling (super-resolution) [18].

7.2 Applications

Notwithstanding the success of the CNNs in Sect. 3, the mere classification of images is of limited practical use, as it gets ambiguous as soon as the image contains more than one dominating object, see the examples in Fig. 6. Locating objects by frames or masks, on the other hand, enables a variety of applications:

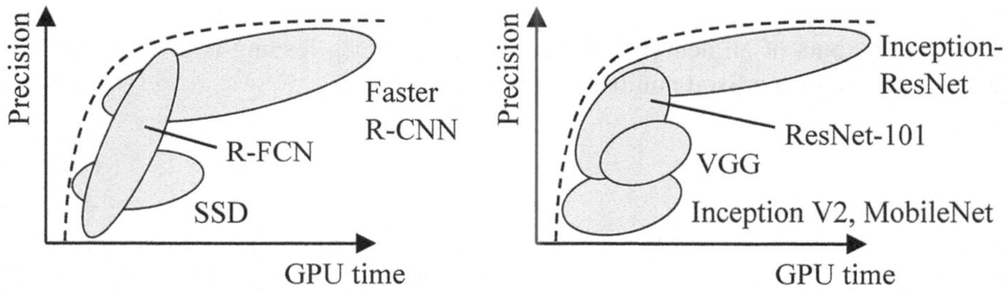

Fig. 13. Precision/speed comparison for object detectors [36] – grouped by meta structure (left) and underlying feature extractor (right)

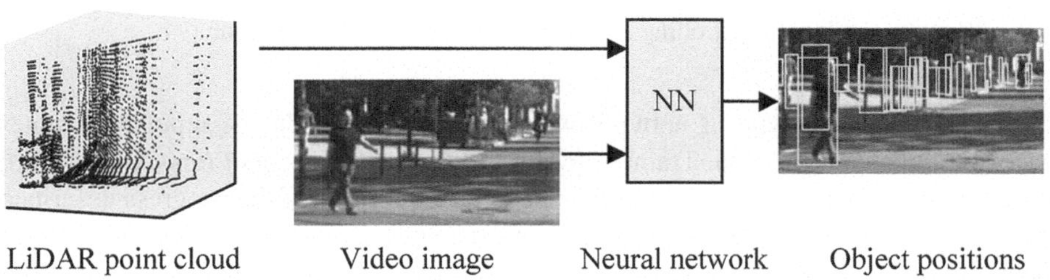

Fig. 14. Fusion of LiDAR and video data for pedestrian detection (images from [56])

- Identifying persons in image collections [21] or video streams.
- Finding anomalies in medical data, such as tomography images [9].
- Detecting prohibited items with airport X-ray scanners.
- Scene analysis for robot movement or automated driving: Detecting pedestrians [2], vehicles [20], traffic signs [97], road markings, or traffic lights [5].

A way for improving the quality of scene understanding is using depth information, e.g., from LiDAR sensors, see Fig. 14. It can be exploited for generating region proposals [56], or serve as an additional image plane. Similarly, information from multiple views can be fused, e.g., from stereo cameras [41], which inspirits the idea of applying CNNs also to volumetric (3D) data.

8 Recurrent Neural Networks (RNNs)

The previous sections were focused on visual perception tasks with CNNs. The second field that substantially advanced with deep learning is any kind of sequence processing, like natural language understanding, or automatic translation.

8.1 Modeling Sequence Data

Sequences are commonly represented by *recurrent* neural networks (RNNs) containing backward connections, other than feed-forward networks. They consist of layers from which a hidden state **h** is fed back to the input, see Fig. 15 – in the simplest case the state is the layer's output **y**. As in (6), boldface face letters denote vectors, i.e., the inputs and outputs of all neurons. The connection to deep leaning is given by unrolling the network over a fixed number of time steps.

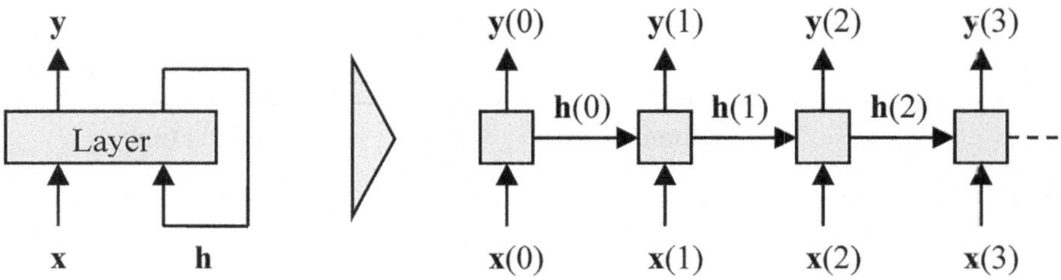

Fig. 15. Unrolling a recursive network layer through time

This is a special type of network which shares the weights of inputs and states over all time steps (Fig. 15). Training by *back-propagation through time* (BPTT) [90] is very similar to standard EBP and thus benefits from most of the progress described in Sects. 4–6, including key factors like dropout [67]. Some quirks need to be considered, e.g., non-saturating activation is still exotic [45], and unrolling is not done explicitly. This notion analogously applies to networks with more than one layer. Depending on the configuration, different types of tasks are possible:

- Sequence-to-vector networks take an input sequence $\mathbf{x}(k)$ and output only the final values $\mathbf{y}(k_{\text{end}})$. Examples are speech recognition or sentiment analysis.
- Vector-to-sequence networks are generators: They produce sequences $\mathbf{y}(k)$ from an initial value $\mathbf{x}(0)$, like in text or speech generation.
- Sequence-to-sequence is the configuration for, e.g., text translation.

Not surprisingly, the most challenging case is training long sequences which correspond to very deep networks. Shortening the sequences or decreasing the sampling rate is often not an option because important information from the past or from finer time scales is lost. One important device for tackling this issue is the use of LSTM cells, which are known for many years [34] but still under research [28, 67, 74].

This "Long Short Term Memory" replaces the recurrent layers of Fig. 15, much like the inception modules in Sect. 3.2 replacing basic convolutional layers. Omitting all details, the LSTM cell keeps an additional long-term state $\mathbf{c}(k)$ and composes its output $\mathbf{y}(k) = \mathbf{h}(k)$ from both states $\mathbf{h}(k-1)$ and $\mathbf{c}(k-1)$. Instead of being engineered, this mechanism is controlled by neural network layers (see Fig. 16) trained from data in an end-to-end fashion, which perfectly aligns with the ideas of deep learning.

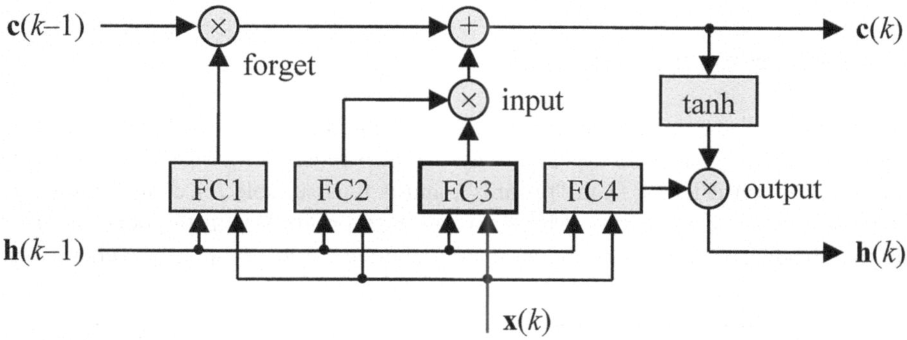

Fig. 16. LSTM cell; FC3 is the main fully connected layer; the other layers are the controllers for input, output and forget gates; output is $\mathbf{y}(k) = \mathbf{h}(k)$

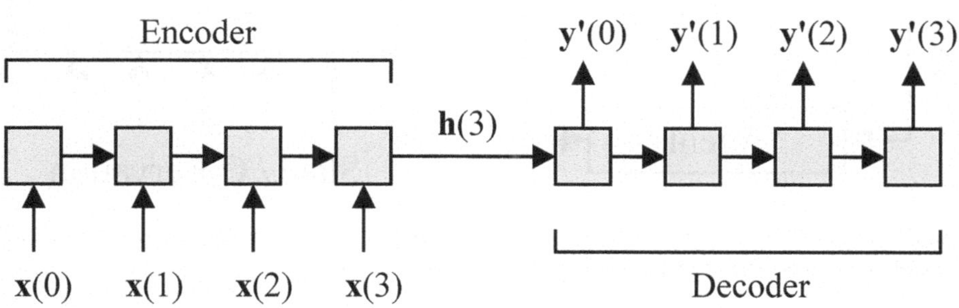

Fig. 17. Simple encoder-decoder network with $\mathbf{h}(3)$ as internal representation

There are also simpler variants like the *Gated Recurrent Unit* (GRU) [10] having a similar performance as LSTM [28]. Independent of the specific type, gated cells are nearly always preferred over plain recurrent layers.

8.2 Applications

Currently, the most publically present application of RNNs is speech recognition [33, 74] in various forms, e.g., voice control in smart phones, cars, or digital assistants (like Apple Siri or Amazon Alexa). Also the quality of automatic text translation has noticeably improved. One key is using *encoder-decoder* networks [10, 83] as shown in Fig. 17: The source text is fed into an encoder producing a state $\mathbf{h}$, followed by a decoder generating the target text from the initialization **h**. Today *attention* mechanisms [3, 55, 94] are established that allow the decoder looking into the source sequence for improving the prediction of the next word.

Encoder-decoder networks are applicable to a variety of tasks by choosing an appropriate encoder. Using a CNN, they can be trained for generating image descriptions [40, 94]. Other possibilities are recognition of handwriting [67], or producing text summaries [77].

9 Reinforcement Learning (RL)

The methods discussed so far use supervised learning on labeled data, possibly accompanied by unsupervised pre-training. *Reinforcement learning* (RL) is different from both in that it learns from "rewards", and is thus suitable for control tasks like robot movement or automated driving. It saw new progress with the debut of deep learning, leading to AlphaGo's victory over one of the world's best Go players [79].

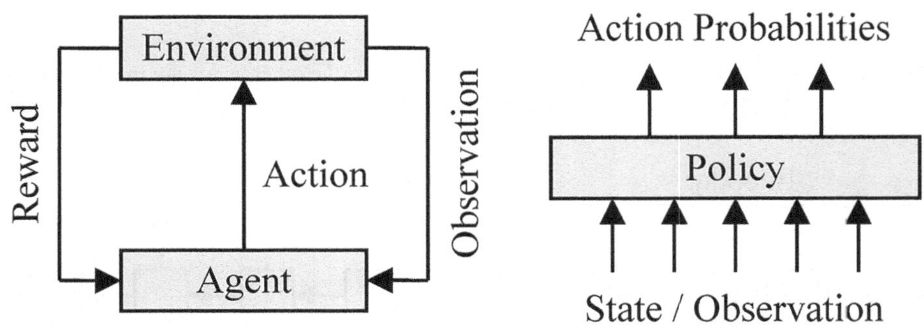

Fig. 18. Principle of reinforcement learning; in the simplest case, the policy takes a state vector as input and outputs the probabilities for a discrete set of actions

RL has its own terminology, illustrated in Fig. 18: An *agent* observes its *environment* and performs *actions*, for which it receives *rewards*. The *policy* is any algorithm that determines the actions, trying to maximize the expected rewards. E.g., a car drives through a scene by steering left/right, capturing its surroundings by video sensors. The reward is the driving time without touching an object.

The impact of deep learning is twofold: First, in form of neural network policies, and second, for estimating the expected rewards from raw data in deep Q-learning.

9.1 Neural Network Policies and Policy Gradients

One approach for learning a policy is to describe it as a function and directly optimize its parameters. This function can be neural network with state s as input and probabilities for all possible actions as output, see Fig. 18. For memory-free systems, the state is simply a representation of the observed environment. Stochastic sampling according to the probabilities ensures that the agent explores new actions. The network parameters are then optimized using policy gradients (PG) as summarized in Algorithm 7, whose concepts originate from the REINFORCE algorithm [91] in 1992.

Alg. 7: Stochastic policy gradients iteration

1. Run the agent for several episodes, usually in a simulated environment.
2. For each experienced state-action pair (s,a), keep the immediate reward r and compute the gradient of w.r.t. the network parameters.
3. Using the data from Step 2, assign to each (s,a) the score $R(s,a)$ computed from all future rewards r_t discounted by the factor $\gamma \in [0,1]$:

$$R(s,a) = \sum_t \gamma^t \cdot r_t \qquad \text{following the path } (s_t, a_t) \tag{7}$$

4. For a "good" score, change the parameters in gradient direction (i.e., make the action more likely), for a "bad" score in negative gradient direction.

The resulting procedure is stochastic with a high gradient variance due to the sparse and delayed rewards, and therefore termed a Monte-Carlo PG method. As it is basically a scaled variant of EBP, Sects. 4–6 apply analogously. This class of algorithms is widely used with numerous improvements addressing variance reduction and extension to continuous action spaces; a good overview is [66]. Research is still in progress [76, 78] and was one source of AlphaGo's success [79].

9.2 Deep Q-learning

A different strategy is *Q-learning*, which basically estimates the expected scores and thereby generates the optimal policy. For this, the system is considered as a Markov decision process (MDP, [6]) with states s, actions a, transition probabilities $T(s, a, s')$ and rewards $r(s, a, s')$, where the triplet (s, a, s') means the transition from s to s' after performing a. The *value* $V(s) = \mathrm{E}[R(s)]$ of a state s denotes the expected discounted reward sum, whereas $Q(s, a) = \mathrm{E}[R(s, a)]$ is the value of s under the assumption that action a is taken. If the Q-values (qualities) for the current state are known, then the optimal policy is taking action a^* having the maximal Q-value.

To understand how Q-values are learned, we start with Bellman's condition [6] for the optimal value $V^*(s)$ of any state s:

$$V^*(s) = \max_a \sum_{s'} T(s, a, s') \left[r(s, a, s') + \gamma \cdot V^*(s') \right] \tag{8}$$

Although its dependence on the next state's optimal value $V^*(s')$ makes it intractable, (9) can be transformed into a *value iteration* [6] that is guaranteed to converge to $V^*(s)$. An equivalent reasoning applies to $Q^*(s, a)$. Since $T(s, a, s')$ and $r(s, a, s')$ are unknown in practice, further reduction leads to TD-(temporal difference)-learning [87] for $V(s)$ and Q-learning for $Q(s, a)$ with learning rate η:

$$V_{k+1}(s) = (1 - \eta) V_k(s) + \eta \left[r + \gamma \cdot V_k(s') \right] \tag{9}$$

$$Q_{k+1}(s, a) = (1 - \eta)Q_k(s, a) + \eta[r + \gamma \cdot \max_{a'} Q_k(s' \cdot a')] \tag{10}$$

Similar to the policy gradients in Sect. 9.1, these are stochastic iterations only on actually seen states and actions. Performing the actions a' can be *off-policy*, i.e., in the extreme case they can be random. In practice, "ε-greedy" algorithms narrow the space more and more to the current policy to balance exploration against execution.

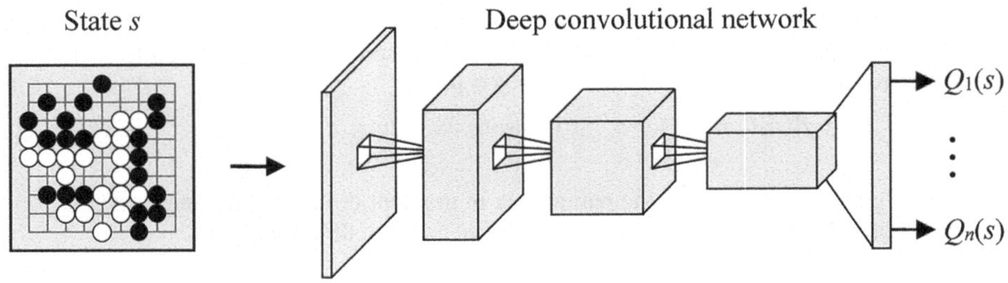

Fig. 19. Deep Q-network estimates Q-values for a finite number of action $a_1, \ldots, a_n$

However, Q-learning in form (9.4) needs an immense number of iterations in setups with many states and actions. Here deep learning comes into play: *Deep Q-networks* (DQNs) are used to approximate $Q(s, a)$; for discrete action spaces, each action can be assigned a value $Q_a(s)$ as shown in Fig. 19. Learning the policy usually runs in an "actor-critic" framework outlined in Algorithm 8.

Alg. 8: Actor-critic framework for approximate deep Q-learning

1. Run the current policy (actor), including a random component (ε-greedy etc.).
2. Generate training data according to (9.4).
3. Train a second DQN (critic) and copy it regularly to the actor DQN.

Compared to policy gradients, the DQN estimates Q-values instead of the probabilities in Fig. 18, and the policy results from training the critic DQN instead of applying gradients. Q-learning is not applicable to all settings but usually converges faster than policy gradients.

9.3 Applications

As pointed out, game-playing is popular field for developing and demonstrating RL algorithms. In fact, AlphaGo [79] is one of the big success stories; it uses policy gradients combined with Monte-Carlo tree search. Value function based methods have some tradition like in TD-Gammon [87], before DQNs had their breakthrough in 2013 [58]: A research group developed an algorithm that learned playing Atari games by approximating the Q-values directly from the screen pixels [59].

An important domain for the use of RL is learning control: Applications are found in robot movement [49, 50] prosthesis control [16], or automated driving [7]; examples are depicted in Fig. 20. Although some results are surprisingly good, it should be kept in mind that these systems learn a control strategy solely from data.

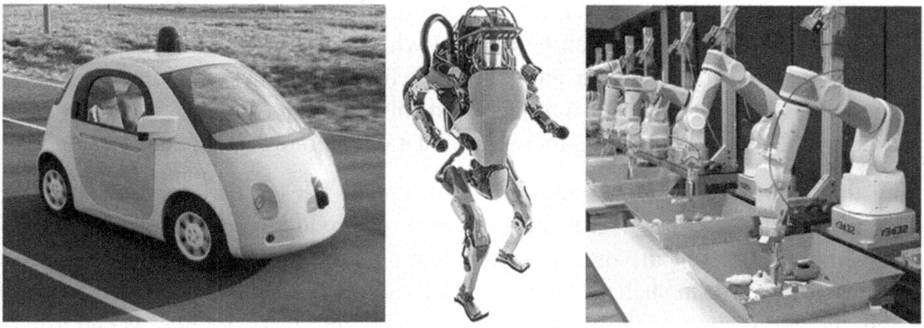

Fig. 20. Control tasks feasible for reinforcement learning: Steering of self-driving cars 88, bipedal locomotion 8, grasping objects with a robot arm 50

A far different RL application is found in [98]: It uses policy gradients for finding optimal neural network structures. This is done for CNNs, competing against the architectures of Sect. 3.2, but also for RNNs aiming to improve LSTM and GRU.

10 Conclusions

This report has covered the tremendous advancements in AI during the past years enabled by deep learning. Training networks with a large number of layers changes an important paradigm: Features are now learned from raw data rather than being engineered by humans. This approach is successful wherever structure needs to be discovered instead of merely approximating functional dependencies. Some of these ideas were known for a long time, but it needed multiple factors coming together:

- Algorithmic improvements (Sect. 4)
- Availability of labeled data and according applications (Sect. 5)
- Computational power, particularly in form of GPUs (Sect. 6.1)
- Frameworks for training and distributing neural networks (Sect. 6.2)

Sometimes the availability of cheap and powerful GPUs is considered as the only reason, but it is debatable whether a breakthrough of this extent could have happened without the other components. Deep learning has impact on mainly three fields:

- CNNs for any kind of visual object detection and localization (Sects. 3 and 7)
- RNNs for language understanding, or text translation (Sect. 8)
- Reinforcement learning for control tasks, automated driving, or games (Sect. 9)

The progress is immense, in some cases heading beyond human performance (image classification, game-playing). It can be observed that the tasks are getting more and more complex or consist of multiple sub-tasks (image captioning etc.) that are learned end-to-end. Since this technology is at the same time accessible to a large community, it must be considered as disruptive:

- It will create new applications, because the quality of the predictions has sizably improved and the methods can be adapted with manageable effort. Tasks that are done by humans today will be automated, potentially with better results.
- The work for developing new products will shift from algorithms to data. Products will include components for gathering user data, preferably in annotated from.

Many things are not a theoretical option but already present in everyday life, like voice control, driver assistance, face recognition, or product recommendations. Complex formations like self-driving cars or autonomously acting robots will show at least a large step forward. In such systems, deep learning will primarily facilitate environmental perception, but also offer tools for learning control. As the field is enduringly progressing, this movement can be expected to continue over the next decade.

For further reading, good overview articles are [48] and [52]. A comprehensive treatment is found in [26], whereas [22] demonstrates the uncomplicated realization. Many individual results are published via arXiv (https://arxiv.org/), corresponding implementations are mostly available at GitHub (https://github.com/).

References

1. AnnotateMyData. http://annotatemydata.com/. Accessed 28 Feb 2018
2. Angelova, A., et al.: Real-time pedestrian detection with deep network cascades. In: British Machine Vision Conference (BMVC), pp. 32.1–32.12 (2015)
3. Bahdanau, D., et al.: Neural machine translation by jointly learning to align and translate. In: 3rd International Conference on Learning Representations (ICLR) (2015)
4. Banko, M., Brill, E.: Scaling to very very large corpora for natural language disambiguation. In: 39th Annual Meeting ot the Association for Computational Linguistics (ACL), pp. 26–33 (2001)
5. Behrendt, K., et al.: A deep learning approach to traffic lights: detection, tracking, and classification. In: IEEE International Conference on Robotics and Automation (ICRA), pp. 1370–1377 (2017)
6. Bellman, R.: A Markovian decision process. J. Math. Mech. **6**(5), 679–684 (1954)
7. Bojarski, M., et al. Explaining how a deep neural network trained with end-to-end learning steers a car. Computing Research Repository, arXiv:1704.07911 (2017)
8. Boston Dynamics: Atlas Robot. https://www.bostondynamics.com/atlas. Accessed 28 Feb 2018
9. Cheng, J., et al.: Computer-aided diagnosis with deep learning architecture: applications to breast lesions in US images and pulmonary nodules in CT scans. Sci. Rep. **6**(24454) (2016)
10. Cho, K., et al.: Learning phrase representations using RNN encoder-decoder for statistical machine translation. In: Conference on Empirical Methods in Natural Language Processing (EMNLP), pp. 1724–1734 (2014)
11. Clevert, D. et al.: Fast and accurate deep network learning by exponential linear units (ELUs). In: 4th International. Conference on Learning Representations (ICLR) (2016)

12. Cordts, M., et al.: The cityscapes dataset for semantic urban scene understanding. In: IEEE Conference on Computer Vision and Pattern Recognition (CVPR), pp. 3213–3223 (2016)
13. CrowdFlower. https://www.crowdflower.com/. Accessed 28 Feb 2018
14. Cybenko, G.: Approximation by superpositions of a sigmoidal function. Math. Control, Signals, Syst. **2**(4), 303–314 (1989)
15. Dai, J., et al.: R-FCN: object detection via region-based fully convolutional networks. In: Advances in Neural Information Processing Systems (NIPS), vol. 29, pp. 379–387 (2016)
16. DiGiovanna, J., et al.: Coadaptive brain-machine interface via reinforcement learning. IEEE Trans. Biomed. Eng. **56**(1), 54–64 (2009)
17. Doersch, C., et al.: Unsupervised visual representation learning by context prediction. In: IEEE International Conference on Computer Vision (ICCV), pp. 1422–1430 (2015)
18. Dong, C., et al.: Learning a deep convolutional network for image super-resolution. In: 13th European Conference on Computer Vision (ECCV), pp. 184–199 (2014)
19. Duchi, J., et al.: Adaptive subgradient methods for online learning and stochastic optimization. J. Mach. Learn. Res. **12**, 2121–2159 (2011)
20. Espinosa, J., et al.: Vehicle detection using AlexNet and Faster R-CNN deep learning models: a comparative study. In: 5th International Visual Informatics Conference (IVIC), pp. 3–15 (2017)
21. Farfade, S., et al.: Multi-view face detection using deep convolutional neural networks. In: 5th ACM on International Conference on Multimedia Retrieval (ICMR), pp. 643–650 (2015)
22. Géron, A.: Hands-On Machine Learning with Scikit-Learn and Tensor-Flow. O'Reilly, Sebastopol (2017)
23. Glorot, X., Bengio, Y.: Understanding the difficulty of training deep feedforward neural networks. In: 13th International Conference on Artificial Intelligence and Statistics, pp. 249–256 (2010)
24. Glorot, X., et al.: Deep sparse rectifier neural networks. In: 14th International Conference on Artificial Intelligence and Statistics, pp. 315–323 (2011)
25. Goodfellow, I., et al.: Generative adversarial nets. In: Advances in Neural Information Processing Systems 27 (NIPS), pp. 2672–2680 (2014)
26. Goodfellow, I., et al.: Deep Learning. MIT Press, Cambridge (2016)
27. Google, Inc.: Neural network processor. Patent WO2016186801 (2016)
28. Greff, K., et al.: LSTM: a search space odyssey. IEEE Trans. Neural Networks Learn. Syst **28**(10), 2222–2232 (2017)
29. He, K. et al.: Delving deep into rectifiers: surpassing human-level performance on ImageNet classification. In: IEEE International Conference on Computer Vision (ICCV), pp. 1026–1034, 2015
30. He, K., et al.: Deep residual learning for image recognition. In: IEEE Conference on Computer Vision and Pattern Recognition (CVPR), pp. 770–778 (2016)
31. He, K., et al.: Mask R-CNN. In: IEEE International Conference on Computer Vision (ICCV), pp. 2980–2988 (2017)
32. Hinton, G., et al.: A fast learning algorithm for deep belief nets. Neural Comput **18**, 1527–1554 (2006)
33. Hinton, G., et al.: Deep neural networks for acoustic modeling in speech recognition: the shared views of four research groups. IEEE Signal Process. Mag. **29**(6), 82–97 (2012)
34. Hochreiter, S., Schmidhuber, J.: Long short-term memory. Neural Comput. **9**(8), 1735–1780 (1997)
35. Hornik, K., et al.: Multilayer feedforward networks are universal approximators. Neural Netw. **2**(5), 359–366 (1989)

36. Huang, J., et al.: Speed/accuracy trade-offs for modern convolutional object detectors. In: IEEE Conference on Computer Vision and Pattern Recognition (CVPR), pp. 3296–3297 (2017)
37. Hubel, D., Wiesel, T.: Receptive fields of single neurones in the cat's striate cortex. J. Physiol. **148**(3), 574–591 (1959)
38. Intel Nervana. https://ai.intel.com/. Accessed 28 Feb 2018
39. Ioffe, S., Szegedy, C.: Batch normalization: accelerating deep network training by reducing internal covariate shift. In: 32nd International Conference on International Conference on Machine Learning (ICML), pp. 448–456 (2015)
40. Karpathy, A., Fei-Fei, L.: Deep visual-semantic alignments for generating image descriptions. IEEE Trans. Pattern Anal. Mach. Intell. **39**(4), 664–676 (2017)
41. Kendall, A., et al.: End-to-end learning of geometry and context for deep stereo regression. In: IEEE Int. Conference on ComputerVision (ICCV), pp. 66–75 (2017)
42. Kingma, D., Ba, J.: Adam: a method for stochastic optimization. In: 3rd International Conference on Learning Representations (ICLR) (2015)
43. Krizhevsky, A. et al.: ImageNet classification with deep convolutional neural networks. In: Advances in Neural Information Processing Systems 25 (NIPS), pp. 1090–1098 (2012)
44. Larsson, G., et al.: Colorization as a proxy task for visual understanding. In: Conference on Computer Vision and Pattern Recognition (CVPR), pp. 840–849 (2017)
45. Le, Q., et al.: A simple way to initialize recurrent networks of rectified linear units. Computing Research Repository, abs/1504.00941 (2015)
46. LeCun, Y., et al.: Handwritten digit recognition with a back-propagation network. In: Advances in Neural Information Processing Systems 2 (NIPS), pp. 396–404 (1990)
47. LeCun, Y., et al.: Gradient-based learning applied to document recognition. Proc. IEEE **86**(11), 2278–2324 (1998)
48. LeCun, Y., et al.: Deep learning. Nature **521**(7553), 436–444 (2015)
49. Levine, S., et al.: End-to-end training of deep visuomotor policies. J. Mach. Learn. Res. **17**(1), 1334–1373 (2016)
50. Levine, S., et al.: Learning hand-eye coordination for robotic grasping with deep learning and large-scale data collection. Int. J. Robo. Res. **37**(4) (2017)
51. Li, Y., et al.: Fully convolutional instance-aware semantic segmentation. In: Conference on Computer Vision and Pattern Recognition (CVPR), pp. 4438–4446 (2017)
52. Littman, M.: Reinforcement learning improves behaviour from evaluative feedback. Nature **521**(7553), 445–451 (2015)
53. Liu, W., et al.: SSD: single shot multibox detector. In: 14th European Conference on Computer Vision (ECCV), pp. 396–404 (2016)
54. Long, J., et al.: Fully convolutional networks for semantic segmentation. In: Conference on Computer Vision and Pattern Recognition (CVPR), pp. 3431–3440 (2015)
55. Luong, M., et al.: Effective approaches to attention-based neural machine translation. Conference on Empirical Methods in Natural Language Processing (EMNLP), pp. 1412–1421 (2015)
56. Matti, D., et al.: Combining LiDAR space clustering and convolutional neural networks for pedestrian detection. In: 14th IEEE International Conference on Advanced Video and Signal Based Surveillance (AVSS), pp. 1–6 (2017)
57. McCulloch, W., Pitts, W.: A logical calculus of the ideas immanent in nervous activity. Bull. Math. Biophys. **5**, 115–133 (1943)
58. Mnih, V., et al.: Playing Atari with deep reinforcement learning. NIPS Deep Learning Workshop (2013)
59. Mnih, V., et al.: Human-level control through deep reinforcement learning. Nature **518**(7540), 529–533 (2015)

60. Nesterov, Y.: A method of solving a convex programming problem with convergence rate $O(1/k^2)$. Sov. Math. Dokl **27**(2), 372–376 (1983)
61. NVIDIA CUDA. https://developer.nvidia.com/cuda. Accessed 28 Feb 2018
62. NVIDIA cuDNN. https://developer.nvidia.com/cudnn. Accessed 28 Feb 2018
63. NVIDIA Jetson. https://developer.nvidia.com/embedded-computing. Accessed 28 Feb 2018
64. NVIDIA Drive. https://developer.nvidia.com/drive. Accessed 28 Feb 2018
65. Pathak, D. et al.: Context encoders: feature learning by inpainting. In: Conference on Computer Vision and Pattern Recognition (CVPR), pp. 2536–2544 (2016)
66. Peters, J., Schaal, S.: Reinforcement learning of motor skills with policy gradients. Neural Netw. **21**(4), 682–697 (2008)
67. Pham, V., et al.: Dropout improves recurrent neural networks for handwriting recognition. In: 14th International Conference on Frontiers in Handwriting Recognition (ICFHR), pp. 285–290 (2014)
68. Polyak, B.: Some methods of speeding up the convergence of iteration methods. USSR Comput. Math. Math. Phys. **4**(5), 1–17 (1964)
69. Redmon, J., et al.: You only look once: unified, real-time object detection. In: Conference on Computer Vision and Pattern Recognition (CVPR), pp. 779–788 (2016)
70. Ren, S., et al.: Faster R-CNN: towards real-time object detection with region proposal networks. In: Advances in Neural Information Processing Systems 28 (NIPS), pp. 91–99 (2015)
71. Rosenblatt, F.: The perceptron: a probabilistic model for information storage and organization in the brain. Psychol. Rev. **65**(6), 386–408 (1958)
72. Rumelhart, D., et al.: Learning representations by back-propagating errors. Nature **323**, 533–536 (1986)
73. Russakovsky, O., et al.: ImageNet large scale visual recognition challenge. Int. J. Comput. Vis. **115**(3), 211–252 (2015)
74. Sak, H., et al.: Long short-term memory recurrent neural network architectures for large scale acoustic modeling. In: 15th Annual Conference of the International Speech Communication Association (INTERSPEECH), pp. 338–342 (2014)
75. Salimans, T., et al.: Improved techniques for training GANs. In: Advances in Neural Information Processing Systems 29 (NIPS), pp. 2234–2242 (2016)
76. Schulman, J., et al.: Trust region policy optimization. In: 32nd International Conference on International Conference on Machine Learning (ICML), pp. 1889–1897 (2015)
77. See, A., et al.: Get to the point: summarization with pointer-generator networks. In: 55th Annual Meeting of the Association for Computational Linguistics (ACL), pp. 1073–1083 (2017)
78. Silver, D., et al.: Deterministic policy gradient algorithms. In: 31st International Conference on International Conference on Machine Learning (ICML), pp. 387–395 (2014)
79. Silver, D., et al.: Mastering the game of Go with deep neural networks and tree search. Nature **529**(7587), 484–489 (2016)
80. Simonyan, K., Zisserman, A.: Very deep convolutional networks for large-scale image recognition. In: 3rd International Conference on Learning Representations (ICLR) (2015)
81. Srivastava, N., et al.: Dropout: a simple way to prevent neural networks from overfitting. J. Mach. Learn. Res **15**, 1929–1958 (2014)
82. Su, H., et al.: Crowdsourcing annotations for visual object detection. In: AAAI Human Computation Workshop, pp. 40–46 (2012)
83. Sutskever, I., et al.: Sequence to sequence learning with neural networks. In: Neural Information Processing Systems 27 (NIPS), pp. 3104–3112 (2014)
84. Szegedy, C., et al.: Going deeper with convolutions. In: Conference on Computer Vision and Pattern Recognition (CVPR), pp. 1–9 (2015)

85. Szegedy, C., et al.: Rethinking the inception architecture for computer vision. In: Conference on Computer Vision and Pattern Recognition (CVPR), pp. 2818–2826 (2016)
86. Szegedy, C., et al.: Inception-v4, Inception-ResNet and the impact of residual connections on learning. In: 31st AAAI Conference on Artificial Intelligence, pp. 4278–4284 (2017)
87. Tesauro, G.: Temporal difference learning and TD-Gammon. Commun. ACM **38**(3), 58–68 (1995)
88. Waymo: Google self-driving car. https://waymo.com/. Accessed 28 Feb 2018
89. Werbos, P.: Beyond regression: new tools for prediction and analysis in the behavioral sciences. PhD thesis, Harvard University (1974)
90. Werbos, P.: Backpropagation through time: what it does and how to do it. Proc. IEEE **78**(10), 1550–1560 (1990)
91. Williams, R.: Simple statistical gradient-following algorithms for connectionist reinforcement learning. Mach. Learn. **8**(3), 229–256 (1992)
92. Wilson, A., et al.: The marginal value of adaptive gradient methods in machine learning. In: Advances in Neural Information Processing Systems 30 (NIPS), pp. 4151–4161 (2017)
93. Xu, B., et al.: Empirical evaluation of rectified activations in convolutional network. In: ICML Deep Learning Workshop, 06–11 July 2015
94. Xu, K., et al.: Show, attend and tell: neural image caption generation with visual attention. In: 32nd International Conference on International Conference on Machine Learning (ICML), pp. 2048–2057 (2015)
95. Zeiler M., Fergus, R.: Visualizing and understanding convolutional networks. In: 13th European Conference on Computer Vision (ECCV), pp. 818–833 (2014)
96. Zhang, Y., et al.: Augmenting supervised neural networks with unsupervised objectives for large-scale image classification. In: 33rd International Conference on International Conference on Machine Learning (ICML), pp. 612–621 (2016)
97. Zhu, Z., et al.: Traffic sign detection and classification in the wild. In: Conference on Computer Vision and Pattern Recognition (CVPR), pp. 2110–2118 (2016)
98. Zoph, B., Le, Q.: Neural architecture search with reinforcement learning. In: 5th International Conference on Learning Representations (ICLR) (2017)

Potential of Virtual Test Environments for the Development of Highly Automated Driving Functions Using Neural Networks

Raphael Pfeffer[1(✉)], Patrick Ukas[2], and Eric Sax[1]

[1] Karlsruhe Institute for Technology, Karlsruhe, Germany
raphael.pfeffer@kit.edu
[2] Hochschule Kempten, Kempten, Germany

Abstract. This paper outlines the implications and challenges that modern algorithms such as neural networks may have on the process of function development for highly automated driving. In this context, an approach is presented how synthetically generated data from a simulation environment can contribute to accelerate and automate the complex process of data acquisition and labeling for these neural networks. A concept of an exemplary implementation is shown and first results of the training of a convolutional neural network using these synthetic data are presented.

1 Introduction

Highly automated driving is currently the key innovating force for the automotive industry. The development and introduction of those functions does not take place abruptly, but evolutionarily through the gradual implementation of driver assistance functions and systems, which show an increasing degree of automation. According to SAE standard J3016, these systems are categorized in six stages, with the highest level, the so-called "Level 5", describing the actual full automation. Level 5 includes a complete automated driving system under all environmental conditions without the need for a human driver to monitor the system or to intervene [1].

An increasingly essential requirement is to ensure the functional safety of these systems in order to guarantee error-free performance. The functional safety of highly automated driver assistance functions can only be ensured by efficient test procedures and adaptation of conventional development processes in industry.

Conventional testing and release processes have economic as well as technical limitations. Thus, with the increasing degree of automation, the number of driving situations to be taken into account also raises. The exponentiality of this connection finally leads to a so-called release trap ("Freigabefalle"). Thus, Wachenfeld and Winner in [2] show that, statistically, approximately 7 billion km of testing would have to be driven on highways in order to be able to release the function "Autobahnpilot" for German highways.

T. Bertram (Hrsg.): *Fahrerassistenzsysteme 2018,* Proceedings, S. 203–211, 2019.
https://doi.org/10.1007/978-3-658-23751-6_18

The gradual introduction of increasingly automated driving functions based on the adoption of already released components is only the first step in meeting this challenge. The goal must be to identify the test cases that are relevant for the release in order to be able to reduce the number of redundant test scenarios and to find alternative test methods that can validly replace real-world test runs or parts thereof.

In the past, simulation-based test methods such as model-in-the-loop (MIL), software-in-the-loop (SIL), hardware-in-the-loop (HIL) or even vehicle-in-the-loop (VIL) [3] have been developed and successfully used to reduce the real-world driving test effort for embedded electronic, mechatronic control devices and control systems [4]. They have become an integral part of the vehicle development process of many OEMs and suppliers. For some systems, such as the Electronic Stability Control (ESC), valid environments have been used in recent years that allow the functional release of these systems for vehicle variants even without the use of real-world test driving (simulation- based homologation [5]). The test cases are derived from the requirements. Those approaches follow the paradigm: "No test without requirement – no requirement without test".

The challenge is different if the requirements can no longer be clearly defined a priori, but rather self-learning functions are introduced. Especially in the context of highly automated driver assistance functions, the application of methods of machine learning and artificial intelligence, which are used in particular for scene interpretation and object recognition, is a key technology. The output quality of these methods, as long as they are based on the principle of neural networks, largely depends on the type and amount of data with which these algorithms are trained. A serious disadvantage is that these training data, if based on real and recorded data, must be annotated manually, which makes this process laborious and time consuming. Approaches as in [6] already show that it is possible in general to train a neural network with a sufficient amount of purely synthetic image data in such a way that it scores at least as well as a net trained purely by real data with respect to object recognition.

The contributions of this paper are:

We will demonstrate an approach to how virtual simulation environments, such as those already available from the established test methods, can be used to automatically train and test neural networks. The aim is to present which potential process improvements regarding the use of neural networks can be achieved by this approach. We will show a possible concept and realization of an environment which allows a neural network to be trained in an automated way. Results of this approach will be presented and compared to reference data.

The paper is organized as follows:

The following chapter deals with related work regarding the training of neural networks based on artificial image data. Section 3 presents the challenges for the development process and identifies the potential for virtual environments. Section 4 shows a concept for automatically generating and listing artificial image data and using them to train neural networks. A possible implementation is presented. Exemplary results using this environment is shown in Sect. 5. A conclusion is drawn in the final chapter.

2 Related Work

In recent years, various approaches and works have contributed to this area, which are briefly presented below:

So far, the focus of many contributions has been on the recognition of objects based on visual data. The concept we will subsequently present is based on an image data approach as well. Works that focus on other types of sensors or consider other approaches (e.g. with focus on end-to-end learning) are not taken into consideration here and must therefore be distinguished from the following.

In the field of pedestrian detection in the area of Advanced Driver Assistance Systems (ADAS), [7] showed how virtual environments can help in learning appearance-based models for pedestrian detection. The authors explored the use of histograms of oriented gradients (HOG) features and linear SVM learning machines to learn a classifier for pedestrian detection. The study showed a rather close performance comparing the detector trained on completely virtual data based on the video game's Half-Life 2 engine on the one hand and on real-world data on the other.

Similar algorithm techniques but with a different data approach were used for example by [8]. In this setup, augmented training samples instead of fully virtual or fully real-world data were used to train a pedestrian classifier. In this work, virtual objects using 19 different pedestrian models were rendered onto a real image background to increase the overall detection performance. The research showed that the classifier trained by exclusively augmented data does not reach the same level of performance as when using a large-scale training based on real data only. However, complementing the real data with augmented data could increase the recognition performance of the pedestrian classifier.

Barbosa et al. investigated in [9] the re-identification rate of individuals using deep learning algorithms in a video surveillance system. In this case, an artificial dataset was generated to increase the amount of training data by using the "MakeHuman" 3-D software and the "Blender" visualization engine. Based on just 50 different artificial 3-D models, a total of over 100,000 instances could be created using variations of different clothes and poses. Benchmarking the deep learning algorithm based on those expanded training data showed a superior performance.

In [10] it was shown that the addition of synthetic generated images to real dataset boosted the detection performance significantly. This setup surpassed both the training variations of the neural network based on purely synthetic data and those based on real data. It is also remarkable that this effect was created with a relatively small amount of real data. In the mixed setup, only 500 real images were used in addition to the 4,000 synthetically generated images. In addition, a decline in accuracy could be observed as the dataset size was increased suggesting overfitting to synthetic data with increase in dataset size.

In the study of [11] the sensitivity of a neural network to different low-level cues in the synthetic training data was investigated. For this purpose 3D pose, foreground, texture and color, background image and color have been varied in different setups. The best results were shown in variants with different views and mixed low level cues.

3 Challenges in the Development Process Context

In science and the automotive industry, there are various approaches to integrating the development of functions based on neural networks into established development processes.

An approach that aims to ensure compliance with previous standards such as Automotive SPICE and ISO 26262 is demonstrated by the W model, which conceptually integrates a V model for data development of neural networks with the standard V model for software development shown in [12].

Taking this into account, we propose a top-level process that is not only used to describe the different stages of function development using neural networks that has to be embedded in the automotive development process but is rather intended to show the hypothetical challenges and potential for virtual environments in the temporal context of a development cycle (Fig. 1).

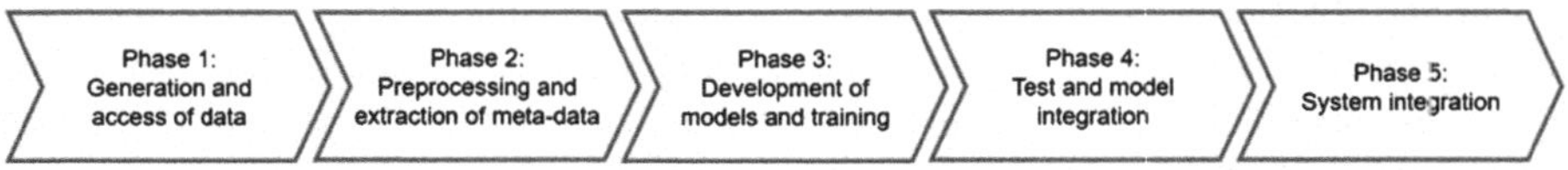

Fig. 1. Neural networks function development process

Phase 1 – Generation and Access of Data. The early availability of the required input data is a major challenge. In addition, this effect is intensified by shorter development processes and the required quantity of data. Recorded data from previous development cycles or predecessor models reach their limits when the new functions cannot be stimulated with the existing available signals (e.g. if they are based on new sensor data). Furthermore, recorded data are dependent on the situations in which they were recorded and can hardly or only with a great deal of effort be adjusted or transferred into situations with different conditions. Synthetic data can be generated with relatively little effort at any time and in almost any quantity.

Phase 2 – Data Processing and Extraction of Meta-data. In order to be able to use the data for neural networks, several steps in preprocessing are necessary. An essential element is the annotation of the data to be used, for example, to train the neural networks. This process step can only be automated with considerable effort and within limits if it is based on real data [13]. In many cases, the subsequent addition of necessary meta-information requires a manual procedure and is therefore cost-intensive and error-prone. Synthetic data generated for this purpose can be provided with all required ground truth data from the outset. In a suitable environment, this step can be fully automated for synthetic data. The annotations generated in this way are fully controllable and error-free.

Phase 3 – Development of Models and Training. One characteristic of algorithms based on neural networks is that their output quality depends largely on the input data used. The functional behavior of a neural network is a combined result of its automatic adaptation through training and its architectural structure. This means that during the development phase of the actual model for a particular function, the adaptability of input data is an essential parameter for the quality of the function to be developed. The environment must ensure that the data in this phase can be flexibly and quickly changed or appropriately recreated (See also the challenges mentioned in phase 1).

Phase 4 – Test, Evaluation and Model Integration. As we have already shown in the introduction to this work, the complete testing of highly automated driving functions is a major challenge. A large number of data driven test cases are required to cover the underlying situation complexity. The transfer of representative test cases from the road to the laboratory must be the goal in order to make testing economically efficient. This challenge is independent of the technological method with which the function was ultimately realized and therefore not subject matter of this work.

Phase 5 – System Integration. During the development and the test of the algorithms, completely different hardware architectures are usually available than in the target system in the vehicle. The challenge for the performance is to provide the algorithms with the identical output behavior for the target system under conditions such as real-time behavior, etc. This is also not the focus of this work.

4 Concept and Technical Approach

Based on the challenges we presented in the previous chapter we will demonstrate a concept which focuses on the potential in phases 1–3. Phases 4 and 5 are not considered in this work.

The aim is to extend an existing simulation environment in a way that all necessary (synthetic) image data can be generated automatically. These data should be used to train a neural network (e.g. an object detection for an ADAS function). In the emulation environment, a large number of virtual driving scenarios should be generated in just a few steps, which should correspond as closely as possible to real scenarios. The image data shall be derived directly from the simulation and the visualization of the generated driving scenarios. For this purpose, it is necessary that all required meta-information is generated during the simulation process as ground truth data or can be derived from the simulation models or quantities. In the example of the object detection, these can be the position data of the target objects in the generated image or sizes of the objects and their classes (e.g. “car”, “truck”, “pedestrian”, “bicycle”). In the preprocessing step, the data generated from the simulation environment are to be prepared in such a way that they can be used as a training data sample for a suitable algorithm. This includes, for example, adapting the data structure, calculative elimination of occlusions in the generated images using the depth information, or sorting out images with target objects that are nonexistent or too small.

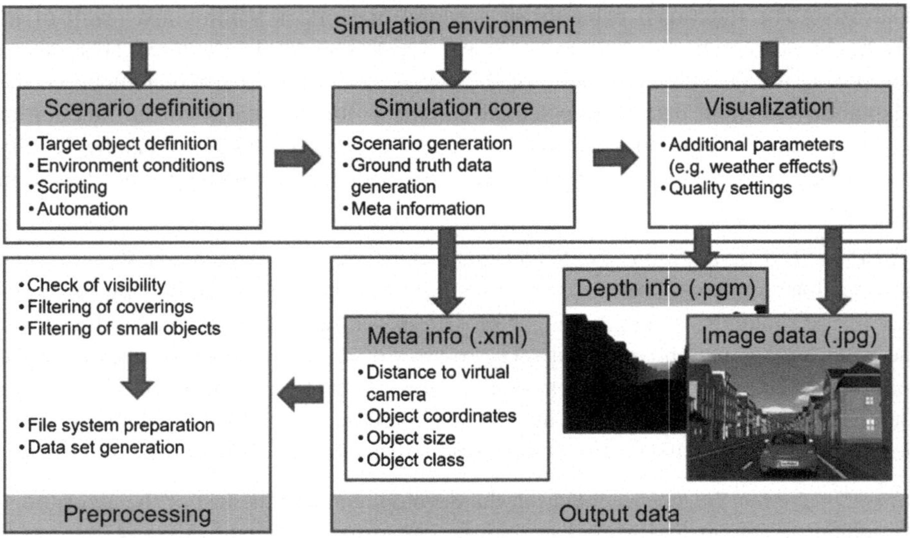

Fig. 2. Concept for automated synthetic data generation with annotations

In our implemented setup, the simulation environment CarMaker (represented by the upper block in Fig. 2) was used for this purpose. In this tool, virtual driving scenarios are generated by an extensive description environment. The driving scenarios of the simulation can be visualized from various perspectives using an integrated animation tool. In addition, a test automation tool is available to generate parameter studies that ultimately allow a variety of scenarios and thus image data to be created. In our approach, script-controlled image data was generated from the visualization, and depth maps were created. In addition, the required meta information were derived via an implemented interface and assigned to the corresponding images. In a subsequent step, the generated data were automatically prepared as described above and stored as a data set (.jpg and .xml). Thus, an environment has been created which is basically able to generate virtually any number of synthetic and annotated data sets in a few steps within a short period of time. The complete environment is shown schematically in Fig. 2.

5 Results

In order to test the implemented setup, 6 different driving scenarios in the simulation environment CarMaker[1] were selected, modified and provided with autonomous traffic. In the simulation phase, approx. 540 images were generated by export from 30 min of simulated scenarios after filtering. The entire driven distance of the simulation is

[1]CarMaker by IPG Automotive GmbH (www.ipg-automotive.com).

about 11 km. In these basic scenarios, the vehicle models of autonomous traffic were replaced by 67 variants. In this way, the database was automatically expanded to approx. 36,000 images with associated meta information. The aim was to train a pre-trained convolutional neural network with the generated data for object detection, in this case the object class "car". For this purpose, a convolutional neural network based on the Single Shot MultiBox Detector (SSD) was selected [14]. This approach is characterized by a moderate accuracy but high speed.

First tests were carried out for this algorithm using the created database. It could be shown that the highest attainable accuracy was achieved by using training data consisting of 10,000 sets. In addition, a strong overfitting effect can be demonstrated from about 7,000 training steps with a batch size of 25. The potential of the synthetic image data could already be demonstrated in this first experiment. The average precision (AP@0.5 PASCAL VOC metric [15]) of the neural network in its original state for the detection of the object class "car" has been improved from 0.3163 to 0.3696 in the peak.

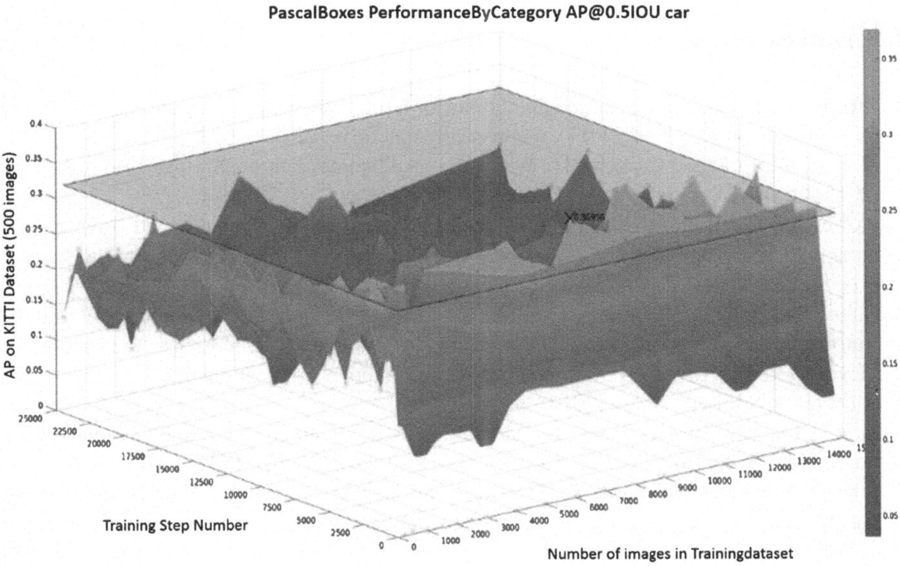

Fig. 3. Average Precision (AP) as a function of training data size and training steps. The red level shows the AP of the SSD in its original state.

This benchmark was created based on a test of 500 real images from the KITTI reference dataset [16]. The AP as a function of training data size and training steps is shown in Fig. 3.

6 Conclusion and Future Works

In this paper we have presented an approach to demonstrate how a possible environment can look like in order to improve the process of data generation and training of neural networks. An already existing simulation environment was used, which was extended in such a way that synthetic and annotated data is generated in a short time and largely automated. We were able to show that the data generated in this way are able to improve the accuracy of an already pre-trained neural network. A real data set for the object recognition of the object class "car" served as a benchmark for this purpose. Furthermore, it was shown that average precision (AP) depends largely on how many virtual data sets are used and how many training steps are done.

Further investigations are to be carried out as a next step. The aim is to analyze the sensitivity regarding different characteristics in the visualization. Various other parameters can be varied for this purpose. On the one hand, parameters such as texture quality and model diversity are conceivable; on the other hand, performance can also be examined with regard to other object classes or other types of neural networks.

References

1. SAE International: Taxonomy and definitions for terms related to driving automation systems for on-road motor vehicles (2017). www.sae.org/autodrive. 22 Feb 2018
2. Maurer, M., Gerdes, J.C., Lenz, B., Winner, H.: Autonomes Fahren. Springer Vieweg, Berlin (2015)
3. Pfeffer, R., Leichsenring, T.: Continuous development of highly automated driving functions with vehicle-in-the-loop using the example of euro NCAP scenarios. In: 7th Conference Simulation and Testing for Vehicle Technology, Berlin (2016)
4. Otten, S., Bach, J., Wohlfahrt, C.,King, C., Lier, J., Schmid, H., Schmerler, S., Sax, E.: Automated assessment and evaluation of digital test drives. In: Zachäus, C., Müller, B., Meyer, G. (eds.) Advanced Microsystems for Automotive Applications 2017. Lecture Notes in Mobility. Springer, Cham (2017)
5. Lutz, A., Schick, B., Holzmann, H.: Simulation methods supporting homologation of Electronic stability control in vehicle variants. Veh. Syst. Dyn. **55**(10), 1432–1497 (2017)
6. Johnson-Roberson, M., Barto, C., Mehta, R., Sridhar, S., Rosaen, K., Vasudevan, R.: Driving in the matrix: can virtual worlds replace human-generated annotations for real world tasks? In: Proceedings of International Conference on Robotics and Automation (ICRA) (2017)
7. Marin, J., Vazquez, D., Geronimo, D., Lopez, A.M.: Learning appearance in virtual scenarios for pedestrian detection. In: IEEE Computer Vision and Pattern Recognition (CVPR) (2010)
8. Nilsson, J., Fredriksson, J., Gu, I.Y.-H., Andersson, P.: Pedestrian detection using augmented training data. In: 22nd International Conference on Pattern Recognition (ICPR) (2014)
9. Barbosa, I., Cristani, M., Caputo, B., Rognhaugen, A., Theoharis, T.: looking beyond appearances: synthetic training data for deep CNNS in re-identification. In: Computer Vision and Pattern Recognition (2017)

10. Rajpura, P.S., Bojinov, H., Hegde, R.S.: Object detection using deep CNNs trained on synthetic images. In: Computer Vision and Pattern Recognition (2017)
11. Peng, X., Sun, B., Ali, K., Saenko, K.: Learning deep object detectors from 3D models. In: Computer Vision and Pattern Recognition (2015)
12. Falcini, F., Lami, G., Constanza, A.: Deep learning in automotive software. In: IEEE Software, May/June 2017, pp. 56–63. IEEE Computer Society (2017)
13. Vondrick, C., Patterson, D., Ramanan, D.: Efficiently scaling up crowdsourced video annotation – a set of best practices for high quality, economical video labeling. Int. J. Comput. Vis. **101**(1), 184–204 (2013)
14. Liu, W., Anguelov, D., Erhan, D., Szegedy, C., Reed, S., Fu, C.-Y., Berg, A.: SSD: Single Shot MultiBox Detector. In: Computer Vision and Pattern Recognition (2015)
15. Everingham, M., Van Gool, L., Williams, C.K.I., Winn, J., Zisserman, A.: The PASCAL Visual Object Classes (VOC) challenge. Int. J. Comput. Vis. **88**(2), 303–338 (2010)
16. Geiger, A., Lenz, P., Urtasun, R.: Are we ready for autonomous driving? The KITTI Vision Benchmark Suite. In: Conference on Computer Vision and Pattern Recognition (CVPR) (2012)

Incorporating Human Driving Data into Simulations and Trajectory Predictions

Manuel Schmidt[1(✉)], Carlo Manna[2], Till Nattermann[2], Karl-Heinz Glander[2], and Torsten Bertram[1]

[1] Institute of Control Theory and Systems Engineering, TU Dortmund University, Dortmund, Germany
manuel.3.schmidt@tu-dortmund.de
[2] ZF Group – TRW Automotive GmbH, Automated Driving and Integral Cognitive Safety, Düsseldorf, Germany

1 Introduction

1.1 Motivation

The development of algorithms for automated driving is a very challenging task. Recent progress in machine learning suggests that many algorithms will have a hybrid structure composed of deterministic or optimization and learning based elements. To train and validate such algorithms, realistic simulations are required. They need to be interaction based, incorporate intelligent surrounding traffic and the other traffic participants behavior has to be probabilistic. Current simulation environments for automotive systems often focus on vehicle dynamics. There are also microscopic traffic simulations that on the other hand don't take vehicle dynamics into account. The few simulation software products that combine both elements still have at least one major problem. That is because lane change trajectories disregard human driving dynamics during such maneuvers. Consequently, machine learning algorithms developed and trained in simulations hardly generalize to non-synthetic data and therefore to real-world applications.

The contribution at hand aims at enhancing simulations through the incorporation of human driving dynamics into lane change maneuver trajectories. Furthermore, the developed module can also be used as a conservative approach to trajectory prediction. This is realized using a one-dimensional Convolutional Neural Network (CNN) [1] as a maneuver classifier together with the sampling of lane change trajectories based on various probability distributions.

1.2 State of the Art

In [2], the problem of lane change trajectory generation is approached using scaled sine curves. Furthermore, polynomials are also often used, because they allow for easy incorporation of boundary conditions, see for example [3].

T. Bertram (Hrsg.): *Fahrerassistenzsysteme 2018*, Proceedings, S. 212–232, 2019.
https://doi.org/10.1007/978-3-658-23751-6_19

For the identification of human driving dynamics during lane changes, the Next Generation Simulation (NGSIM) Dataset is used [4]. The authors of [5] analyzed this dataset from the perspective of microscopic traffic simulations.

However, to the knowledge of the authors, currently, no publication is available that tries to identify human driving dynamics during lane change maneuvers based on the NGSIM dataset with a focus on trajectory generation. Our approach uses just a few hundred real trajectories and can be applied easily in the automotive industry. In contrast, other manifold learning approaches like Generative Adversarial Nets need thousands of data samples for the training to model the dynamics.

2 Analysis of the Data

The NGSIM Dataset was created and recorded in the United States of America. Several cameras were mounted above different highway-stretches. Through the combination of object detection, tracking and lane association, a very rich dataset was obtained that also includes many lane change maneuvers. Such an exemplary maneuver is shown in Fig. 1. Every trajectory of the tracked traffic participants is described in a global coordinate frame. This makes the dataset unique. Most available datasets reference in a moving coordinate system, like the ego-vehicle coordinate system. Such data first needs to be transformed into a stationary coordinate frame using vehicle odometry data that is prone to errors.

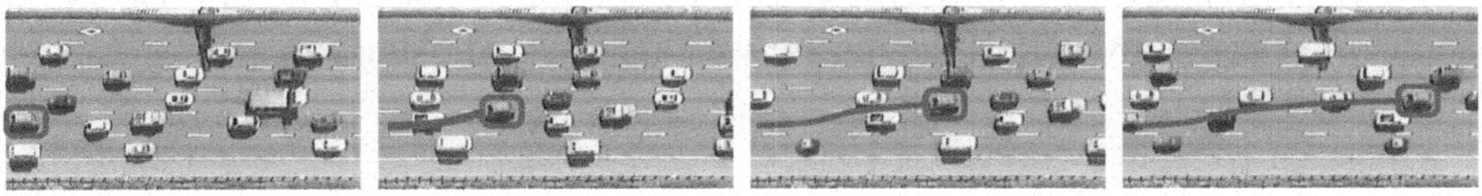

Fig. 1. Exemplary lane change in the NGSIM dataset.

To identify human driving dynamics during lane changes, the lane change trajectories are analyzed in detail. A stationary curvilinear coordinate system is used for this purpose. It allows handling all trajectories the same way. The first step consists in filtering the data using a Butterworth Filter [6] with a cutoff frequency $f_c = 1.5$ Hz. To ensure that this operation does not result in phase delay, forward-backward filtering is used. The extraction of lane change trajectories is done in several steps. Since traffic participants are assigned to lanes on the highways, the time t_T of transition to the destination lane is identified. Afterwards, the trajectories:

$$X(t), t \in [t_\text{T} - 10\,\text{s}\; t_\text{T} + 10\,\text{s}], \tag{1}$$

$$Y(t), t \in [t_\text{T} - 10\,\text{s}\; t_\text{T} + 10\,\text{s}], \tag{2}$$

are extracted in the interval $[t_\text{T} - 10\,\text{s}\; t_\text{T} + 10\,\text{s}]$. Then a manual inspection of the obtained trajectories and their derivatives $\dot{X}(t)$, $\dot{Y}(t)$ is conducted to sort out lane changes that show visible signs of tracking errors. This way, a total of 302 lane changes are identified. In the last step, the start and end of the lane change is manually identified and the data saved for further processing. This data serves as the starting point for the human driving dynamics analysis and modeling. Two quantities are of high importance for trajectory generation, the lane change duration T_{LC} and the lateral lane deviation defined as:

$$d_{\mathrm{dev}} = \frac{(d - d_{\mathrm{LW}})}{d_{\mathrm{LW}}}. \tag{3}$$

Figure 2 illustrates the notation that is used in this contribution.

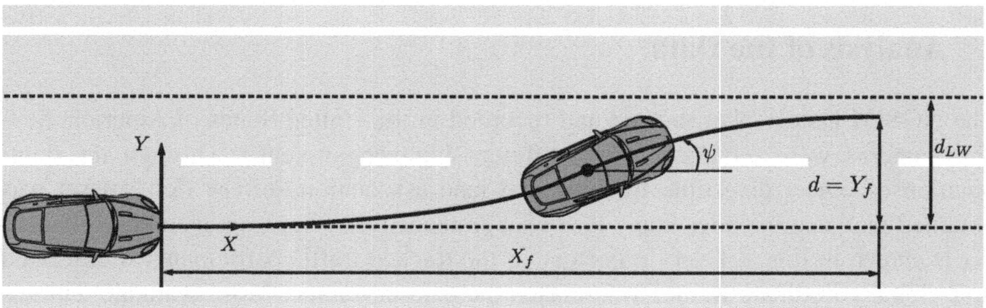

Fig. 2. Illustrates the notation that is used in this contribution.

Using a Kolmogorov Smirnoff Test [7], we identified that both empirical distributions represented by its histograms are well represented using generalized gamma distributions [8]:

$$p(x, a, b, c) = \frac{\frac{c}{a^b} x^{b-1} e^{-\left(\frac{x}{a}\right)^c}}{\Gamma\left(\frac{b}{c}\right)}, \tag{4}$$

with hyperparameters a, b, c and the data x. The empirical distributions and fitted models are show in Figs. 3 and 4.

We are furthermore interested if the lane change duration T_{LC} depends on the mean of the longitudinal velocity prior to the maneuver. For this, the mean of the longitudinal velocity over a horizon of $t = 1.5$ s before the start of the lane change is used. The data is shown in Fig. 5. It is well represented using a Gaussian Mixture Model (GMM) [9]:

$$\begin{aligned} \hat{p}\left(\mathbf{x} = \left(T_{LC}, \bar{v}_{\mathrm{lon},[t=-1.5\,\mathrm{s}\,t=0\,\mathrm{s}]}\right)\right) &= \sum\nolimits_{\mathrm{i}=1}^{K} \lambda_{\mathrm{i}} \cdot \frac{1}{2\pi |\Sigma_{\mathbf{i}}|^{\frac{1}{2}}} \mathrm{e}^{-\frac{1}{2}(\mathbf{x}-\boldsymbol{\mu}_{\mathbf{i}})^{\mathrm{T}} \Sigma_{\mathrm{i}} (\mathbf{x}-\boldsymbol{\mu}_{\mathbf{i}})} \\ &= \sum\nolimits_{\mathrm{i}=1}^{K} \lambda_{\mathrm{i}} \cdot \mathrm{Norm}_{\mathbf{x}}(\boldsymbol{\mu}_{\mathbf{i}}, \boldsymbol{\Sigma}_{\mathbf{i}}), \end{aligned} \tag{5}$$

with mean vectors $\boldsymbol{\mu}_{\mathbf{i}} \in \mathbb{R}^2$, covariance matrices $\boldsymbol{\Sigma}_{\mathbf{i}} \in \mathbb{R}^{2\times2}$ and weights $\lambda_{\mathrm{i}} \in \mathbb{R}$.

Using the Bayesian Information Criterion (BIC) [10] for model selection, it is found that a GMM with two components and diagonal covariance matrices serves as a good model to represent the data. Inspection of Fig. 5 suggests that the lane change duration is approximately statistically independent of the mean of the longitudinal velocity before the lane change. We come to this conclusion because the parametric model results in a probability distribution that is approximately parallel to the velocity axis.

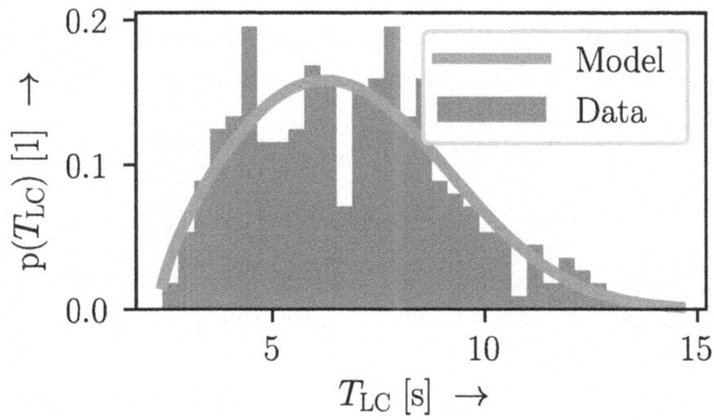

Fig. 3. Histogram and model of lane change duration data.

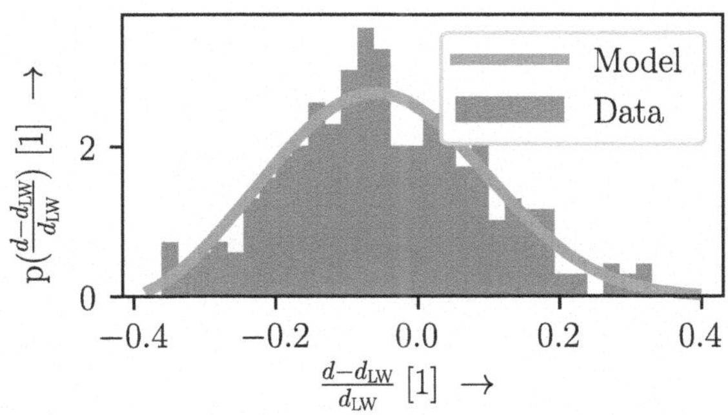

Fig. 4. Histogram and model of lateral lane deviation data.

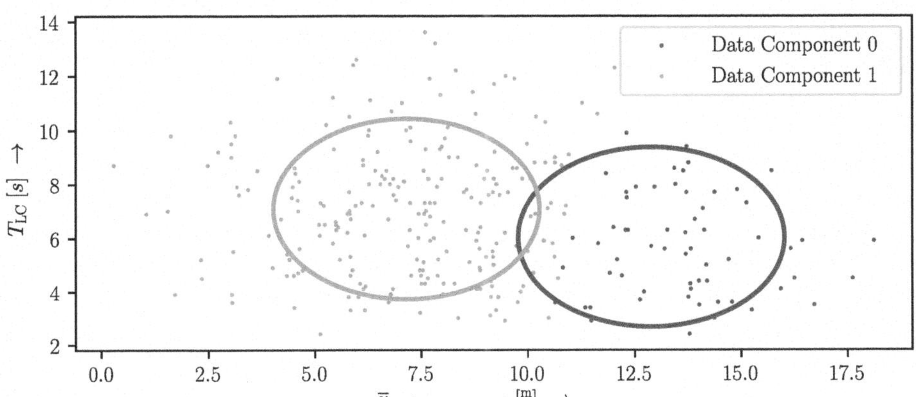

Fig. 5. Lane change duration over mean of longitudinal velocity prior to lane change. The two fitted GMMs are shown and the colours visualize the data clustering.

3 Development of a Module for the Generation of Naturalistic Lane Change Trajectories

3.1 Mathematical Background of Polynomial Trajectory Generation

We assume that lane change trajectories can be represented using third order polynomials both for $X_P(t)$ and $Y_P(t)$. Their parametric form is as follows:

$$X_P(t) = a_3 t^3 + a_2 t^2 + a_1 t + a_0, \tag{6}$$

$$Y_P(t) = b_3 t^3 + b_2 t^2 + b_1 t + b_0. \tag{7}$$

The subscript P denotes polynomials. A total of four boundary conditions for each trajectory are necessary to solve for the unknown parameters $\{a_3, a_2, a_1, a_0\}$ and $\{b_3, b_2, b_1, b_0\}$.

Depending on the task, the following two sets of boundary conditions are used:

Variant 1:

$X_P(0) = X_0$	$\dot{X}_P(0) = v_0 \cdot \cos \psi_0$	$X_P(T_{LC}) = X_f$	$\dot{X}_P(T_{LC}) = v_f \cdot \cos \psi_f$
$Y_P(0) = Y_0$	$\dot{Y}_P(0) = v_0 \cdot \sin \psi_0$	$Y_P(T_{LC}) = Y_f$	$\dot{Y}_P(T_{LC}) = v_f \cdot \sin \psi_f$

Variant 2:

$X_P(0) = X_0$	$\dot{X}_P(0) = v_0 \cdot \cos \psi_0$	$\dot{X}_P(T_{LC}) = v_f \cdot \cos \psi_f$	$\ddot{X}_P(T_{LC}) = 0$
$Y_P(0) = Y_0$	$\dot{Y}_P(0) = v_0 \cdot \sin \psi_0$	$Y_P(T_{LC}) = Y_f$	$\dot{Y}_P(T_{LC}) = v_f \cdot \sin \psi_f$

Herein, ψ denotes the heading angle of the vehicle. Variant 1 is useful in cases of lane changes in which the target position on the neighboring lane is constrained and lies in a small goal area. On the other hand, Variant 2 doesn't place a condition on the total traveled distance in the longitudinal direction during the lane change. This is especially useful if one aims at the conservative prediction of lane change trajectories. Both variants lead to a linear system of equations of the form:

$$\mathbf{A} \cdot \mathbf{p} = \mathbf{x}_B, \tag{8}$$

with the parameter vector $\mathbf{p} \in \mathbb{R}^4$, the boundary condition vector $\mathbf{x}_B \in \mathbb{R}^4$ and the matrix $\mathbf{A} \in \mathbb{R}^{4\times4}$ that relates both.

Higher order polynomials could also be used if there are even more constraints that should be imposed on the trajectories, for example, acceleration and jerk constraints. The focus in the following is on third order polynomials.

3.2 Incorporation of Human Driving Dynamics using Frequency-Domain Methods

From a theoretical probabilistic perspective, the task of incorporation of human driving dynamics is mathematically precise defined. Lane change trajectories are samples of the following generative probability distribution:

$$p(X(t), Y(t)|\mathbf{X}_\mathrm{S}), \tag{9}$$

in which $\mathbf{X}_\mathrm{S} \in \mathbb{R}^m$ stands for a m-dimensional vector that represents the world state. This vector includes features of the traffic situation, the dynamics of the vehicle under consideration, properties of the driver and many more attributes. This formulation however is not very useful for practical applications since the dimension m is not known and most probably very high dimensional. Research in the field of modelling such high dimensional probability distributions is just in the beginning phase, see for example [11] that deals with Generative Adversarial Neural Networks that aim at the modelling of generative probability distributions. Such neural networks need thousands of trainings samples to learn the underlying generative distribution. The NGSIM dataset on the other hand only provides only a few hundred suitable trajectories.

We therefore take another approach. As described, the approach for modeling the data is not generative. Instead, the data is analysed and the frequency content of difference trajectories $Y_\mathrm{D}(t) = Y(t) - Y_\mathrm{P}(t)$ is modeled with the goal in mind to represent the data distribution of the 302 real trajectories better as by using just polynomial trajectories. The boundary conditions for $Y_\mathrm{P}(t)$ are taken from the real trajectory $Y(t)$ under consideration. Figure 6 illustrates the approach that is going to be described in the following.

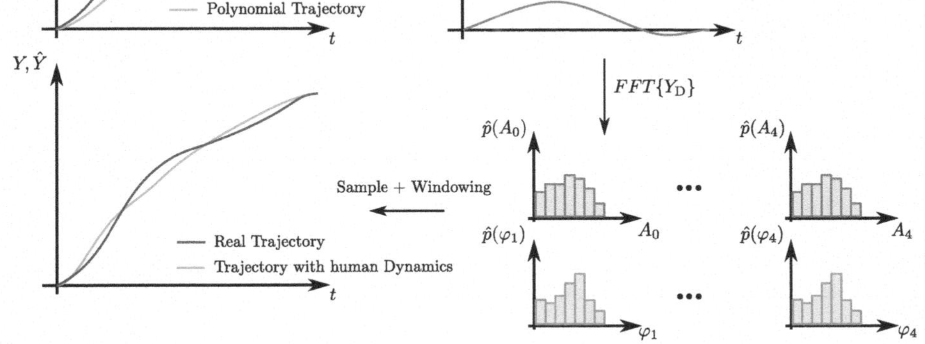

Fig. 6. Overview of our approach to incorporate human driving dynamics into trajectory generation. Using the frequency spectrum of the lateral difference trajectory, a joint probability distribution is used to model the amplitudes and phases of the offset and first four harmonics. Finally, sampling from this distribution leads together with some post-processing to trajectories with incorporated human dynamics.

Our focus is on the dynamics of the lateral movement. However, a similar approach can be used for the longitudinal movement, additionally taking into account that there is often a trend of increasing velocities in case of lane changes with the goal of overtaking. The starting point of the analysis lies in the assumption that polynomial trajectories for the lateral movement $Y_P(t)$ of a traffic participant serve as a suitable prior. Based on this assumption, for every of the 302 lane change trajectories in the dataset, we extract the boundary conditions consisting of the initial heading $\psi(t = 0) = \psi_0$, the initial lateral velocity $\dot{Y}(t = 0)$ and the traveled lateral distance $Y(t = T_{LC})$ are extracted. The last boundary condition needed for calculating the coefficients of a third order polynomial $Y_P(t)$ is the final lateral velocity $\dot{Y}(t = T_{LC})$. Figure 9 shows one exemplary lane change trajectory together with the polynomial fit and the difference trajectory $Y(t) - Y_P(t)$.

A Fast Fourier Transform (FFT) is applied to all 302 lateral difference trajectories $Y_D(t)$ to obtain the frequency spectrum. Zero-Padding has to be used because the trajectories have different durations. Padding with zeros ensures that the frequency resolution is the same for all transformed signals. Afterwards, looking at the amplitudes of the discrete spectrum, for every transformed difference trajectory, the corresponding frequencies of the first five highest amplitudes are collected. Figure 7 shows a histogram of these frequencies. The analysis of the histogram suggests, that in total five harmonics have to be considered to reconstruct the frequency content of the lateral difference trajectories. Our experiments suggest that focusing only on four harmonics is still sufficient and makes the task of modelling the probability distribution easier. The histogram of all four harmonics and their phase are shown in Figs. 10, 11, 12, 13 and 14.

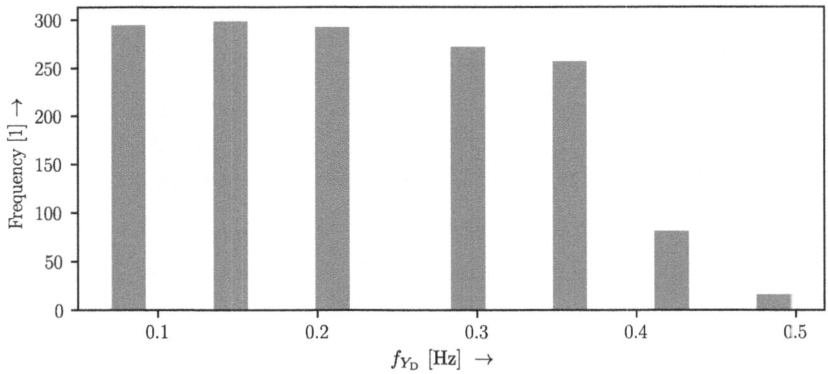

Fig. 7. Histogram of frequency components that correspond to the five highest amplitudes in the discrete Fourier Spectrum of all difference trajectories.

Our approach to incorporate human driving dynamics consists of three fundamental steps. First, data analysis is conducted to develop insights into the data. Based on the observations, we aim to fit a parametric model to represent the probability distribution a part of the Fourier spectrum:

$$p(A_0, A_1, \varphi_1, A_2, \varphi_2, A_3, \varphi_3, A_4, \varphi_4) \tag{10}$$

using a nine-dimensional Gaussian Mixture Model. The Bayesian Information Criterion is used for model selection purposes. Figure 8 shows that the data is best represented using seven components with full covariance matrices. The fit is based on the Expectation Maximization algorithm [12].

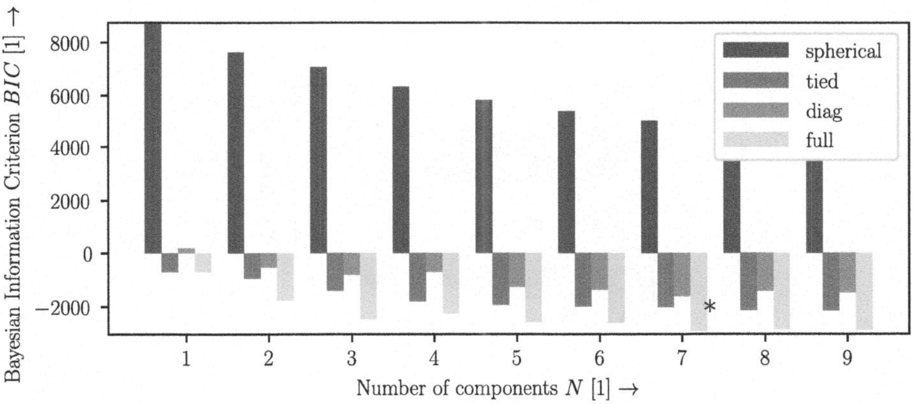

Fig. 8. Model selection of Gaussian Mixture Model for the synthesis of difference trajectories based on the Bayesian Information Criterion. Different colours represent different forms of covariance matrices. The asterisk highlights the chosen model.

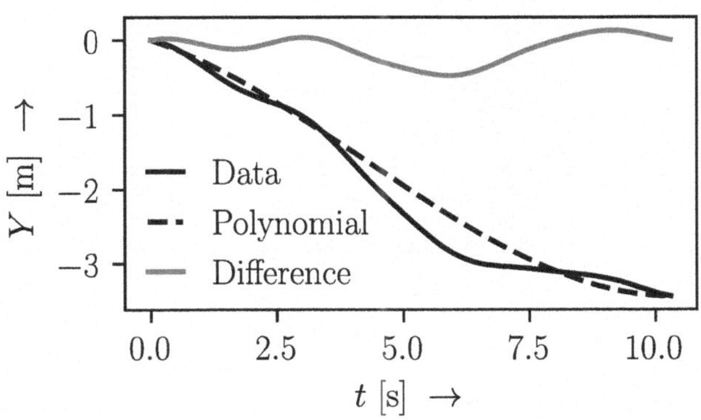

Fig. 9. Comparison of real trajectory, polynomial and difference trajectory.

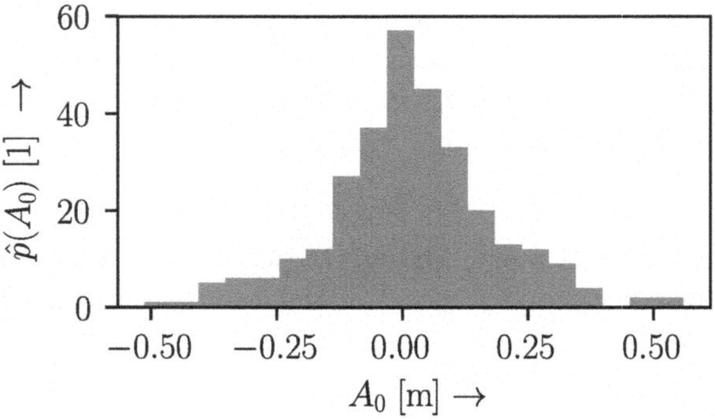

Fig. 10. Histogram of the offset A_0.

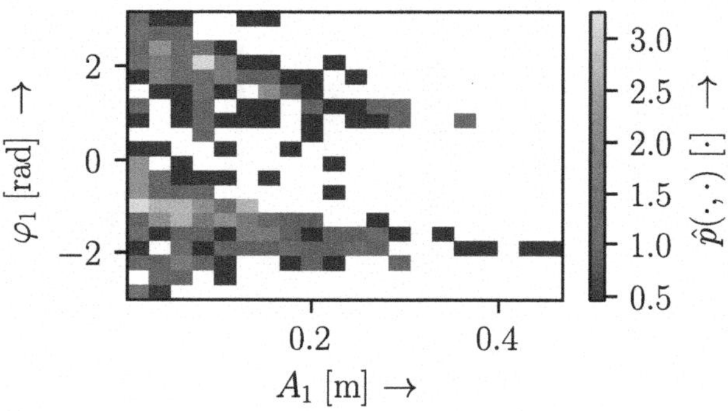

Fig. 11. Histogram of amplitude and phase of first harmonic.

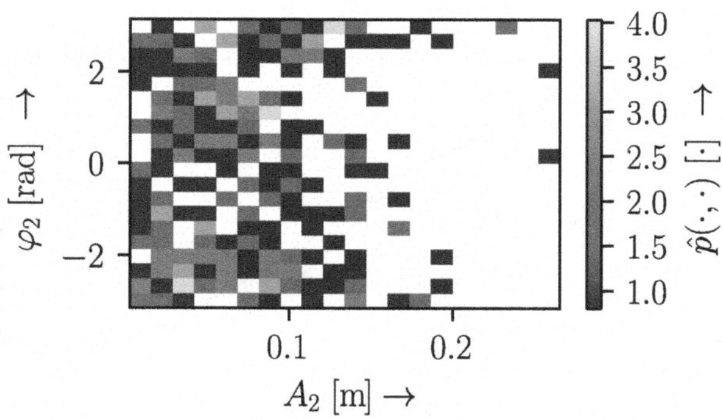

Fig. 12. Histogram of amplitude and phase of second harmonic.

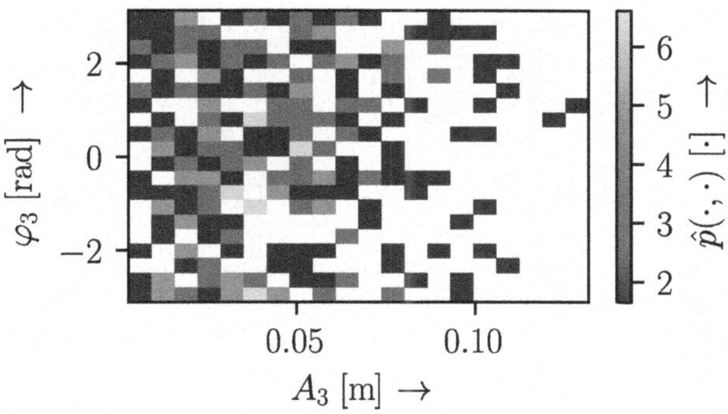

Fig. 13. Histogram of amplitude and phase of third harmonic.

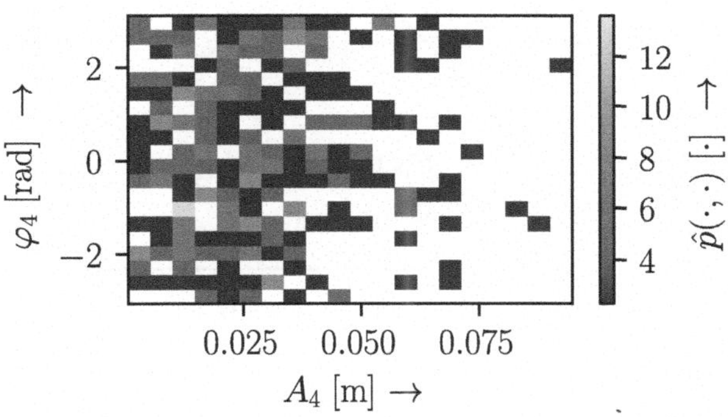

Fig. 14. Histogram of amplitude and phase of fourth harmonic.

The synthesis of a lateral lane change trajectory is done in two steps. First, the amplitudes A_{i} and phases φ_{i} are sampled from the Gaussian Mixture Model. Finally, the trajectory is calculated using:

$$\widehat{Y}(t) = Y_{\mathrm{P}}(t) + \widehat{Y}_{\mathrm{D}}(t) = Y_{\mathrm{P}}(t) + \widehat{A}_0 + \sum\nolimits_{\mathrm{i}=1}^{4} \widehat{A}_{\mathrm{i}} \cdot \cos\left(2\pi\omega_{\mathrm{i}} \cdot t + \widehat{\varphi}_{\mathrm{i}}\right), \quad (11)$$

with the sampled offset $\widehat{A}_0$, the sampled amplitudes of the harmonics $\widehat{A}_{\mathrm{i}}$ and their corresponding phases $\widehat{\varphi}_{\mathrm{i}}$.

One challenge that arises through this approach is that the boundary conditions that are placed during the synthesis of the polynomial trajectory are in the general case violated after addition of a synthesized difference trajectory $\widehat{Y}_{\mathrm{D}}(t)$.

The polynomial fulfills the boundary conditions so that it is necessary that the difference trajectory is zero at the beginning and the end. That ensures that the velocity boundary conditions are satisfied. Furthermore, the slope of the difference signal also needs to be zero at both ends. The blue signal in Fig. 15 shows a sampled difference trajectory. To fulfill the boundary conditions, a discrete Tukey window [13] is used.

Its mathematical description in discrete time is as follows:

$$w(n) = \begin{cases} \frac{1}{2}\left[1 + \cos\left(\pi\left(\frac{2n}{\alpha(N-1)} - 1\right)\right)\right], & 0 \leq n \leq \frac{\alpha(N-1)}{2} \\ 1, & \frac{\alpha(N-1)}{2} \leq n \leq (N-1)\left(1 - \frac{\alpha}{2}\right) \\ \frac{1}{2}\left[1 + \cos\left(\pi\left(\frac{2n}{\alpha(N-1)} - \frac{2}{\alpha} + 1\right)\right)\right], & (N-1)\left(1 - \frac{\alpha}{2}\right) \leq n \leq (N-1) \end{cases} \tag{12}$$

with the discrete time index n, the total number of samples N and a hyperparameter α. We choose this parameter as $\alpha = 0.75$. Three Tukey Windows with different α are plotted over normalized time t_N in Fig. 16. Applying a correctly scaled window to the sampled difference trajectory results in satisfaction of the boundary conditions, see the orange curve in Fig. 15.

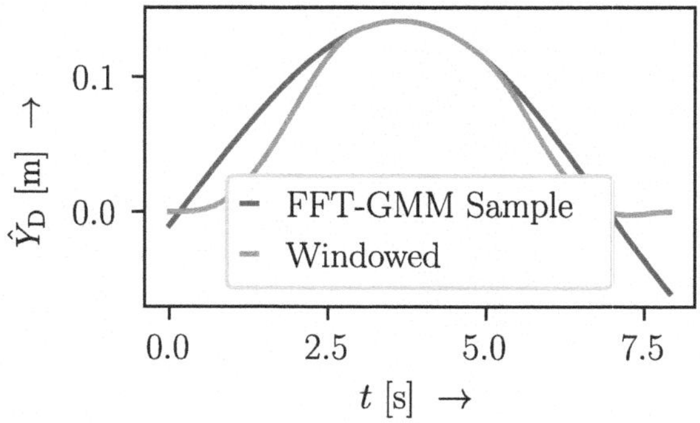

Fig. 15. Synthesized difference trajectory with and without windowing.

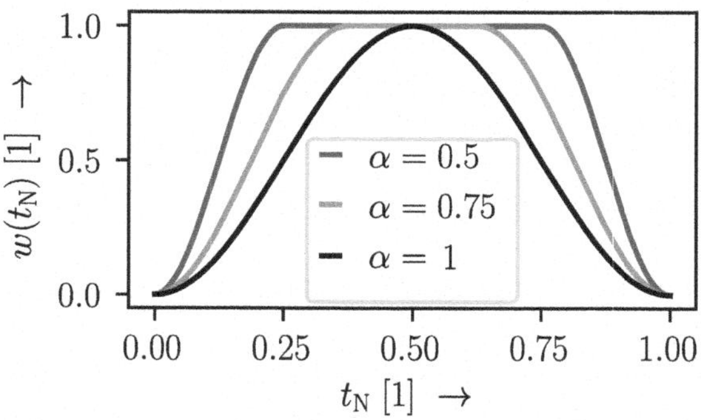

Fig. 16. Tukey windows for different values of α.

Comparison of trajectories is shown in Fig. 17. The effect of incorporation of difference trajectories appears as higher variance around the polynomial trajectory which serves as the mean.

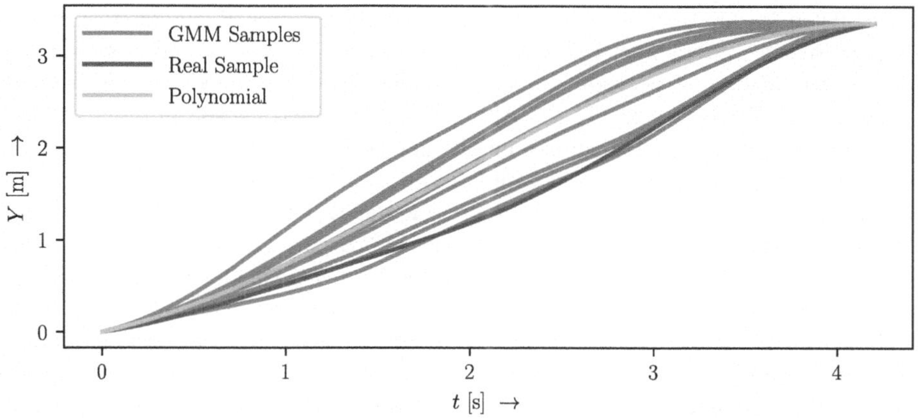

Fig. 17. Comparison of a real trajectory, the polynomial fit and ten with incorporated dynamics using GMM samples.

4 Experimental Results

The evaluation would be obvious if a generative model for direct sampling of lane change trajectories was used. A suitable metric like the Kullback-Leibler Divergence [14] or Wasserstein Distance [15] could be chosen to measure the similarity of real distribution and modeled distribution.

Since our approach is different and non-generative, the evaluation is more difficult. The idea is to compare the data distributions of the real trajectories to the ones with incorporated driving dynamics through using polynomials in combination with difference trajectories.

The evaluation is based on the histograms of $Y(t), Y_P(t)$ and $Y_S(t) = \widehat{Y}(t) = Y_P(t) + \widehat{Y}_D(t)$ at discrete times $\{T_1 := t = 1\,\mathrm{s}, \ldots, T_8 := t = 8\,\mathrm{s}\}$. The subscript P represents polynomial trajectories and *S* sampled trajectories from the final model that uses polynomials and synthesized difference trajectories. Empirical distributions for T_3 are shown in Figs. 18, 19 and 20. They reveal, that the incorporation of driving dynamics leads to a spread of the data distribution compared to the case when only a polynomial is used. This in turn leads to better approximation of the real data distribution. Parametric Gaussian Mixture Models are fitted to the data for better illustration. The model selection is again based on the Bayesian Information Criterion.

One way to measure the dissimilarity of probability distributions is by using the Kullback-Leibler Divergence and it is defined as follows for two discrete distributions:

$$D(p||q) = KL(p, q) = \sum\nolimits_x p(x) \cdot \log \frac{p(x)}{q(x)} \tag{13}$$

The term

$$D_{KL}\big(R, S/P\big) := KL(R, S) - KL(R, P), \tag{14}$$

is calculated for the time slices $T_1, \ldots T_8$ and the abbreviation R denotes real trajectories, P denotes polynomial trajectories and S stands for sampled trajectories based on the combination of polynomials and sampled difference trajectories. Results of the evaluation are shown in Fig. 21. Above time range was chosen in order to have enough samples for the calculation of empirical distributions. Since the Kullback-Leibler Divergence is a measure for dissimilarity of two distributions, the plot in Fig. 21 reveals that the sampled trajectories represent the real trajectories better because the value is negative for every time slice. The improvement is less strong for times $t \leq 1$ s because the sampled difference trajectories play a smaller role there and so the trajectory is similar to the polynomial trajectory. One reason for this is because of the windowing that is used to maintain the fulfilment of the boundary conditions. Most improvements happen for time slices T_2 up to T_5 and are a direct consequence of the higher variance represented in the sampled trajectories compared to polynomials. Afterwards, the effect decreases slowly because for most lane changes, the lane change duration T_{LC} is shorter than 8 s such that the windowing reduces the variance in the data again. Summarizing, the improvement using human dynamics is very significant and shows the advantages of the application of the developed module over polynomial trajectories alone for lateral movement.

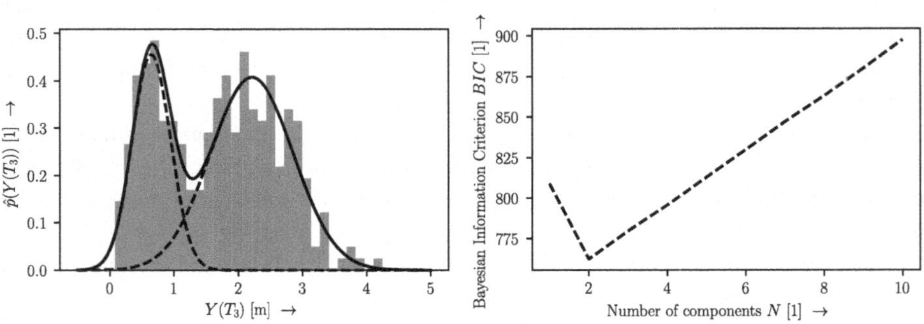

Fig. 18. Histogram of fitted GMM and BIC curve for time T_3 of all real trajectories.

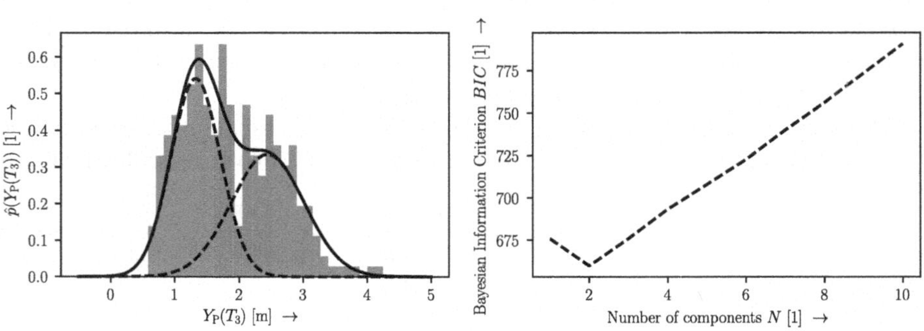

Fig. 19. Histogram of fitted GMM and BIC curve for time T_3 of all polynomial trajectories.

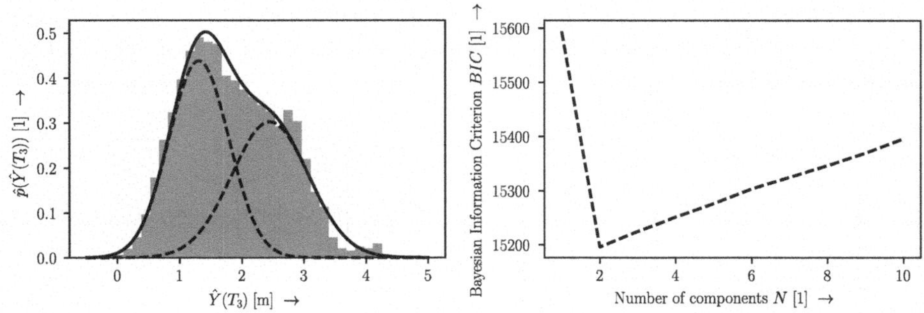

Fig. 20. Histogram of fitted GMM and BIC curve for time T_3 of trajectories with incorporated dynamics.

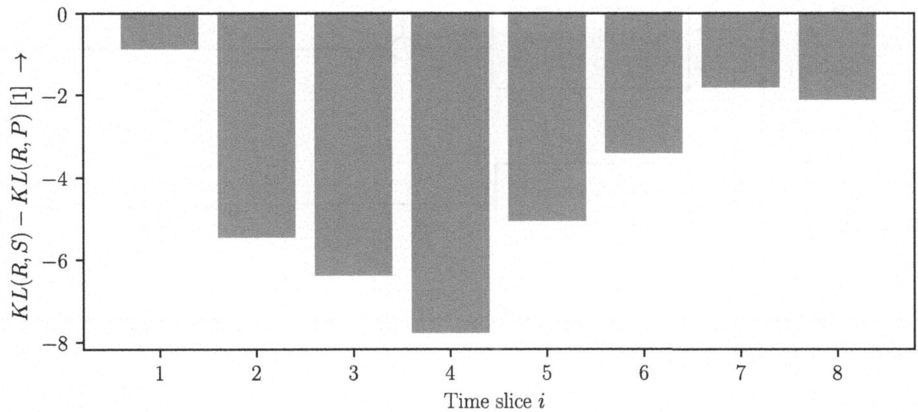

Fig. 21. Difference of Kullback-Leibler Divergences for all time slices under consideration.

5 Applications of the Developed Module

5.1 Usage in Simulations

One application of the developed module is the generation of lane change trajectories in simulations. Such can be constrained where several conditions on the trajectories are imposed. On the other hand, if the traffic scene is not dense, also free lane changes can occur which have more free parameters, like the duration of the lane change.

The typical process of the synthesis of a lane change trajectory for a free lane change is visualized in Fig. 22. The lane change duration T_{LC}, the lateral lane deviation d_{dev} and the frequency domain characteristics of the difference trajectory $\{A_0, A_1, \varphi_1, A_2, \varphi_2, A_3, \varphi_3, A_4, \varphi_4\}$ are sampled. The $X(t)$ and $Y(t)$ components are handled separately.

Automotive simulation products often use a hierarchical approach for lane changes of traffic vehicles. We therefore assume, that there exists a signal that represents the intention of a lane change to one of the adjacent lanes. The developed module is triggered by the lane change intention signal. Finally, the generated trajectories $\widehat{X}(t)$ and $\widehat{Y}(t)$ are sent to the traffic vehicle under consideration, checked for collisions and executed.

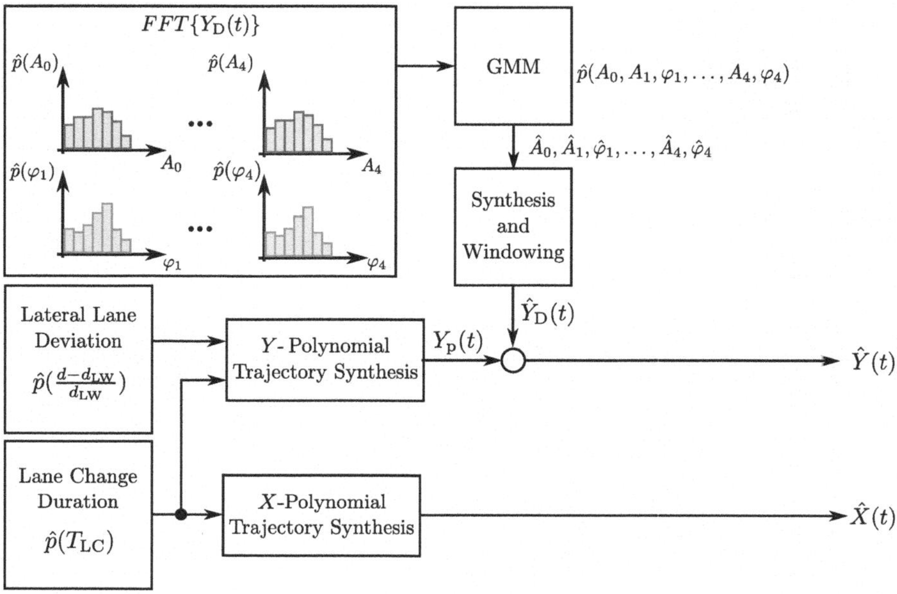

Fig. 22. Illustration of one possible application of the developed modules for the generation of free lane change trajectories and conservative trajectory predictions.

5.2 Hierarchical Trajectory Prediction based on Maneuver Classification

The second application scenario is the usage for trajectory predictions. Once it is obvious that a traffic participant is going to change lane, sampling of many trajectories gives a conservative lane change corridor in the spatiotemporal *X*-*Y*-*t* domain that the vehicle is likely going to be within during the lane change.

Trajectory prediction can be subdivided into two subtasks. First, based on several features like for example the lateral velocity or lateral distance to the neighboring lane marker of a traffic participant, a maneuver classification is done. The task is mathematically defined as:

$$\widehat{cls} = \arg\max_{cls} p(cls|\mathbf{O}) \tag{15}$$

There are three possible classes *LCL* (Lane Change Left), *LCR* (Lane Change Right) and *LK* (Lane Keeping), $cls \in \{LCL, LCR, LK\}$. The observation $\mathbf{O}$ consists of a total of 15 feature vectors that describe a vehicle that we want to predict a lane change trajectory for, $\mathbf{O} = [\boldsymbol{f}_1 \boldsymbol{f}_2 \cdots \boldsymbol{f}_{15}]$. These fifteen features include for example the lateral distance of the considered traffic participant to the adjacent lane markings and various other descriptive features. Every feature vector represents one second history of a specific discrete feature signal $f_1(t)$ with a stepsize of $\Delta T = 0.1$ s,

$f_m athbf1(k) = f_1(k \cdot \Delta T)$. The data was collected using vehicles equipped with a front-facing camera and radar sensors. Labelling was done manually. A detailed description of the dataset and the used features can be found in [16]. A novel method for the maneuver classification based on one dimensional Convolutional Neural Network is proposed in this contribution. The motivation for this is based on the idea that the feature signals show specific trends that can be detected easily with such Convolutional Neural Networks. Most machine learning approaches to maneuver classification flatten all features and therefore loose the temporal coherencies. The proposed architecture of the Neural Network is shown in Fig. 23. Table 1 shows the parameters and network layers that are chosen. A ReLU (Rectified Linear Unit) is a specific type of activation function for CNNs and was introduced in [17]. The last architectural part of the network consists of a dense, cross entropy and softmax layer and transforms the features into categorical class proper probabilities.

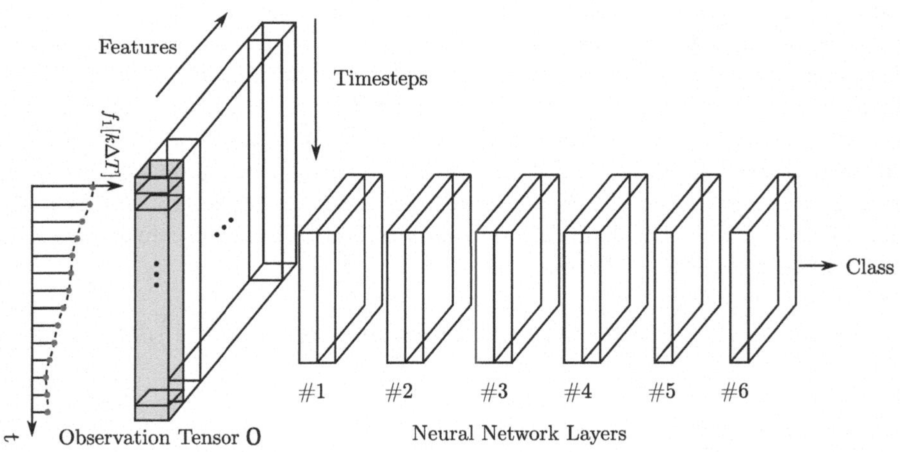

Fig. 23. Convolutional Neural Network architecture for the maneuver classification.

Table 1. Architectural elements of 1D-CNN maneuver classifier

#1	1D-Convolution (filters: 30, kernel size: 3) + ReLU + Batch Normalization
#2	1D-Convolution (filters: 60, kernel size: 2) + ReLU + Batch Normalization
#3	1D-Convolution (filters: 120, kernel size: 2) + ReLU + Batch Normalization
#4	1D-Convolution (filters: 160, kernel size: 2) + ReLU + Batch Normalization
#5	Flatten Layer
#6	Dense Layer + Cross Entropy + Softmax

We compare the results of the 1D-CNN architecture to a state of the art gradient boosted tree classifier, XGBoost [18]. The dataset description is shown in Table 2. Since the dataset is highly unbalanced, we choose to oversample to balance it. Afterwards, the dataset is split into training, validation and test-set, refer to Table 3.

Table 2. Dataset characteristics

	LCR	LCL	LK
Unbalanced	4230	1950	294284
Balanced (Oversampling)	20358	20154	48063

Table 3. Train/Validation/Test split of the dataset

	LCR	LCL	LK
Train	20000	20000	20000
Validation	195	100	13993
Test	163	54	14070

Confusion matrices of both classifiers on unseen test data are shown in Figs. 24 and 25. The comparison reveals that the performance of the 1D-CNN is consistently better for lane changes whereas XGBoost performs better for lane keeping. We conclude that it would be beneficial to use both in form of an ensemble of experts. After the maneuver classification and obtaining $p(cls|f_1(t),\ldots,f_{15}(t))$, the trajectory prediction is realized using sampling based on the probability distributions for the lane change duration $\widehat{p}(T_{\mathrm{LC}})$, the lateral lane deviation $\widehat{p}(d_{\mathrm{dev}})$ and the frequency components $\widehat{p}(A_0,A_1,\varphi_1,A_2,\varphi_2,A_3,\varphi_3,A_4,\varphi_4)$ for the difference trajectory synthesis. The final longitudinal velocity boundary condition is probabilistically enhanced using a truncated Gaussian Normal Distribution:

$$v_{\mathrm{X,f}} = v_{\mathrm{X}}(T_{\mathrm{LC}}) = v_{\mathrm{X,0}} + \max\left(v_{\mathrm{X,D}} \sim \mathrm{Norm}\left(\mu = v_{\mathrm{X,0}}, \sigma^2 = 0.1\right), 0.8 \cdot v_{\mathrm{X,0}}\right), \tag{16}$$

with mean μ and standard deviation σ^2.

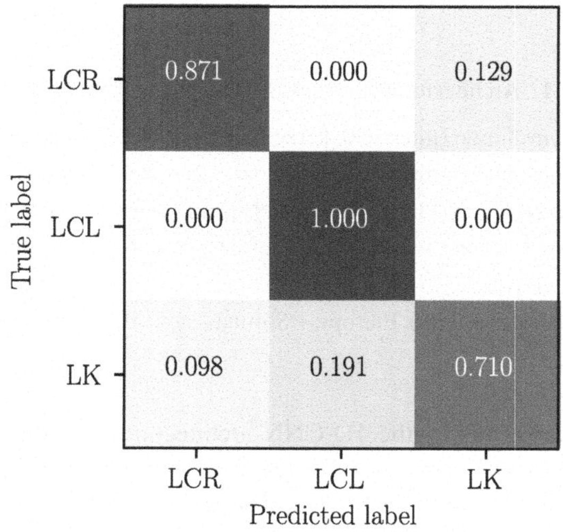

Fig. 24. Confusion matrix of 1D-Convolutional Neural Network maneuver classifier.

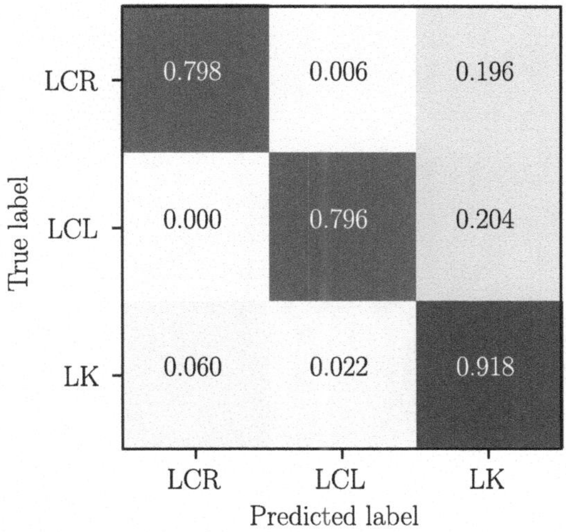

Fig. 25. Confusion matrix of XGBoost maneuver classifier.

The following probability is assigned to every predicted trajectory

$$p\left(\widehat{X}(t), \widehat{Y}(t)\right) \propto \frac{p(cls|f_1(t), \ldots, f_{15}(t)) \cdot \widehat{p}(T_{\mathrm{LC}}) \cdot \widehat{p}(d_{\mathrm{dev}})}{c_N}, \tag{17}$$

with a normalization constant c_{N}.

Finally, Fig. 26 shows a total of 20 samples trajectories in the spatiotemporal X-Y-t domain. The probabilistic nature of the sampling results in trajectories that span a volume in the spatiotemporal domain and every trajectory has an assigned probability. This predictions can be used in decision making and motion planning algorithms to ensure safety. Since of the uncertain movement of a traffic participant during a lane change, a conservative approach to decision making and motion planning results in total avoidance of the spanned volume. On the other hand the assigned probabilities are of high importance in critical situation for collision avoidance systems. It might be beneficial sometimes to accept a low risk for collision with another traffic participant if a collision with a static object in the environment can be avoided that would have worse impacts on the ego-vehicles passengers.

Figure 27 illustrates the projection of another set of sampled trajectories onto the X-Y plane. This again shows that the samples differ much if compared to each other – a beneficial feature for the use in conservative and safe automated driving architectures.

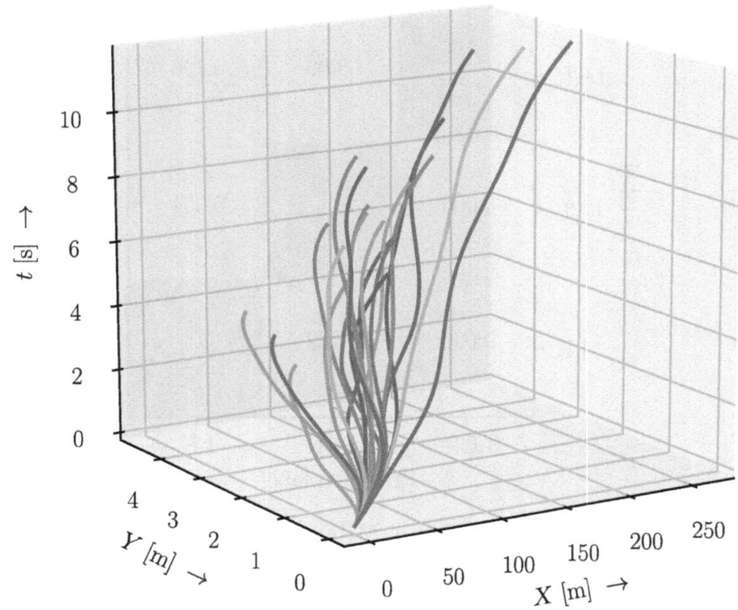

Fig. 26. Sampleyectories in spatiotemporal X-Y-t domain.

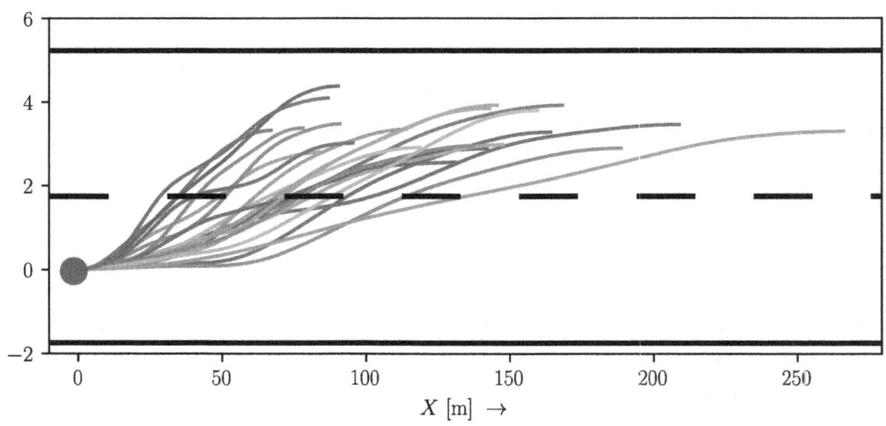

Fig. 27. Another set of trajectory samples projected on the X-Y-plane.

6 Conclusion

Automated vehicles are going to rely heavily on machine learning for perception and decision making algorithms. The development, training and validation of such algorithms are done in simulations. This demands very realistic simulations. The contribution shows a practical and efficient way of incorporating human driving dynamics into automotive simulations and furthermore into trajectory predictions. Our module is very flexible, has various interfaces based on various boundary conditions that can be imposed on lane change trajectories and it makes direct use of parametric models of probability distributions that reflect human driving during lane change maneuvers. Extensions can be incorporated easily since the module is programmed using object orientation.

It is planned to integrate the module in the simulation software that is used in the Driving Simulator of the Institute of Control Theory and Systems Engineering of TU Dortmund University. This enables us to evaluate the model subjectively and finetune it with the goal of enhancing the use of simulations for machine learning. Our evaluation based on the Kullback-Leibler Divergence shows that the described approach achieves higher similarity of the data distributions compared to the use of polynomial trajectories alone. But so far, no statement can be made whether the synthesized trajectories also feel natural to a human driver. This is left for future work.

References

1. Lecun, Y., Bottou, L., Bengio, Y., Haffner, P.: Gradient-based learning applied to document recognition. Proc. IEEE **86**(11), 2278–2324 (1998)
2. Schreier, M.: Bayesian environment representation, prediction, and criticality assessment for driver assistance systems. Dissertation (2016)
3. You, F., Zhang, R., Lie, G., Wang, H., Wang, H., Xu, J.: Trajectory planning and tracking control for autonomous lane change maneuver based on the cooperative vehicle infrastructure system. Expert Syst. Appl. **42**(14), 5932–5946 (2015)
4. US Department of Transportation: NGSIM – next generation simulation (2007). http://www.ngsim.fhwa.dot.gov
5. Thiemann, C., Treiber, M., Kesting, A.: Estimating acceleration and lane-changing dynamics based on NGSIM trajectory data. Transp. Res. Rec. J. Transp. Res. Board. **2088** (2008)
6. Butterworth, S.: On the theory of filter amplifiers. Exp. Wirel. Wirel. Eng. **7**, 536–541 (1930)
7. Massey, F.J.: The Kolmogorov-Smirnov test for goodness of fit. J. Am. Stat. Assoc. **46**(253), 68–78 (1951)
8. Stacy, E.: A generalization of the gamma distribution. Ann. Math. Stat. **33**, 1187–1192 (1962)
9. Prince, S.: Computer Vision: Models Learning and Inference. Cambridge University Press (2012)
10. Schwarz, G.: Estimating the dimension of a model. Ann. Stat. **6**, 461–464 (1978)
11. Goodfellow, I., et al.: Generative adversarial nets. Adv. Neural Inf. Process. Syst. **27**, 2672–2680 (2014)
12. Dempster, A., Laird, N., Rubin, D.: Maximum likelihood from incomplete data via the EM algorithm. J. Roy. Stat. Soc. B (Methodol.) **39**,1–38 (1977)
13. Tukey, J.: An introduction to the calculations of numerical spectrum analysis. In: Spectral Analysis of Time Series, pp. 25–46 (1967)
14. Kullback, S., Leibler, R.: On information and sufficiency. Ann. Math. Stat. **22**(1), 79–86 (1951)
15. Vallender, S.: Calculation of the Wasserstein distance between probability distributions on the line. Theor. Prob. Appl. **18**(4), 784–786 (1972)

16. Krüger, M., Stockem, N.A., Nattermann, T., Glander, K.H., Bertram, T.: Lane change prediction using neural networks considering classwise non-uniformly distributed data. In: Proceedings of the 9th GMM-Symposium AmE 2018 – Automotive meets Electronics (2018)
17. Krizhevsky, A., Sutskever, I., Hinton, G.: Imagenet classification with deep convolutional neural networks. Adv. Neural Inf. Process. Syst. **25**,1097–1105 (2012)
18. Chen, T., Guestrin, C.: XGBoost: a scalable tree boosting system. In: KDD 2016 Proceedings of the 22nd ACM SIGKDD International Conference on Knowledge Discovery and Data Mining, pp. 785–794 (2016)

Deep Learning-Based Multi-scale Multi-object Detection and Classification for Autonomous Driving

Maximilian Fink[1(✉)], Ying Liu[1], Armin Engstle[1], and Stefan-Alexander Schneider[2]

[1] AVL Software and Functions GmbH, Regensburg, Germany
maximilian.fink@avl.com
[2] University of Applied Sciences, Kempten, Germany

Abstract. Autonomous driving vehicles need to perceive their immediate environment in order to detect other traffic participants such as vehicles or pedestrians. Vision based functionality using camera images have been widely investigated because of the low sensor price and the detailed information they provide. Conventional computer vision techniques are based on hand-engineered features. Due to the very complex environmental conditions this limited feature representations fail to uniquely identify a specific object. Thanks to the rapid development of processing power (especially GPUs), advanced software frameworks and the availability of large image datasets, Convolutional Neural Networks (CNN) have distinguished themselves by scoring the best on populthis information, the boundingar object detection benchmarks in the research community. Using deep architectures of CNN with many layers, they are able to extract both low-level and high-level features from images by skipping the feature design procedures of conventional computer vision approaches. In this work, an end-to-end learning pipeline for multi-object detection based on one existing CNN architecture, namely Single Shot MultiBox Detector (SSD) [1], with real-time capability, is first reviewed. The SSD detector predicts the object's position based on feature maps of different resolution together with a default set of bounding boxes. Using the SSD architecture as a starting point, this work focuses on training a single CNN to achieve high detection accuracy for vehicles and pedestrians computed in real time. Since vehicles and pedestrians have different sizes, shapes and poses, independent NNs are normally trained to perform the two detection tasks. It is thus very challenging to train one NN to learn the multi-scale detection ability. The contribution of this work can be summarized as follows:

- A detailed investigation on different public datasets (e.g., KITTI [2], Caltech [3] and Udacity [4] datasets). The datasets provide annotated images from real world traffic scenarios containing objects of vehicles and pedestrians.
- A data augmentation and weighting scheme is proposed to tackle the problem of class imbalance in the datasets to enable the training for both classes in a balanced manner.
- Specific default bounding box design for small objects and further data augmentation techniques to balance the number of objects in different scales.

T. Bertram (Hrsg.): *Fahrerassistenzsysteme 2018*, Proceedings, S. 233–242, 2019.
https://doi.org/10.1007/978-3-658-23751-6_20

- Extended SSD+ and SSD^2 architectures are proposed in order to improve the detection performance and keeping the computational requirements low.

Keywords: Autonomous driving · Environment perception and understanding · Artificial intelligence · Deep learning · Feature learning and extraction · Vehicle detection · Pedestrian detection · Data augmentation

1 Introduction

The automotive industry is undergoing huge change. With the vision of future mobility, almost all automotive companies together with their suppliers are investing in the development of intelligent vehicles. According to the European iMobility Forum for Vehicle Automation, the vehicles of the future will provide safe, smart and clean mobility with zero accidents, zero delays as well as no negative impact on the environment such as noise or carbon emissions [5].

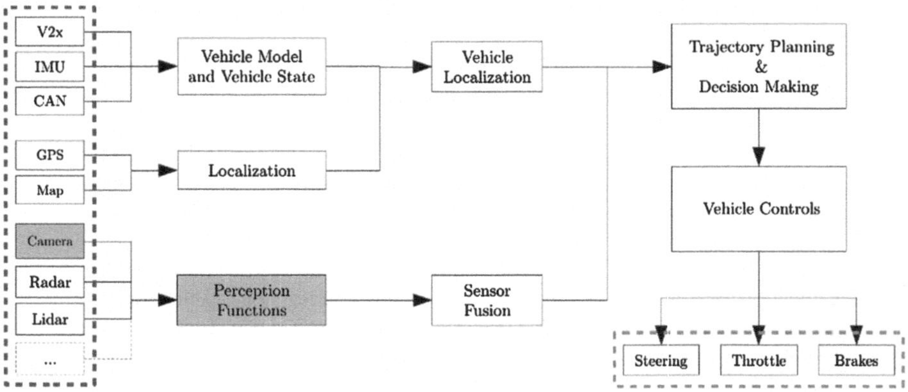

Fig. 1. End-to-end system for autonomous driving from AVL Software and Functions GmbH.

A complete function chain of a system for autonomous driving vehicles is shown in Fig. 1. The input stack acts as a sensory system imitating the human senses, such as eyes or the sense of equilibrium. The inputs coupled with the perception block give the car vital information about its surroundings. Together with the localization of the vehicle, a simple (up to SAE level 3) or more advanced trajectory planning and decision making (SAE level 4 and 5) is possible. By having a desired path and drive request, the system controls the longitudinal and lateral movement and triggers throttle, brakes and steering as needed to perform the decided manoeuvre.

Perceiving the environment of the car is one key functionality to enable autonomous diving cars. Since cameras are available at relatively low cost compared with other sensors such as RADAR or LiDAR, the implementation of a lot of functionality is attempted based on image information. A front view camera is often used for various detection tasks, such as object, traffic sign and lane detection. Conventional computer vision methodologies can be used to extract the features, which represent the marking within the image. Those approaches use the fact that the lines feature high contrast compared to the road pavement. Making use of this characteristic, gradient based algorithms, where filters slide over all image pixels, can extract edges from the image. This hand-engineered feature design, which refers to the features subtracted using the conventional computer vision approach, such as edges and corners, is sensitive to noise and it completely ignores the shape (long, thin structure) of the line markings. Furthermore, edges appear everywhere in the image and this may lead to false extraction results. With the rapid development in faster processing units, advanced machine learning technologies have come into the focus of the automotive industry. In particular, deep neural networks are dominating conventional computer vision approaches for perception function tasks in various benchmarks. By using multiple, hidden layers, the deep architectures of CNN are able to extract both low-level and high-level features from images by skipping the feature design procedures of conventional computer vision approaches. This performance comes with a demand for high computational resources, which makes deep learning algorithms not always real-time capable, even on the most state-of-the-art hardware. This observation motivates this work and the aim was to implement a deep learning based detection algorithm that can predict the 2D position of vehicles and pedestrians based on images from a single camera system cameras at a high detection speed.

2 Convolutional Neural Networks

In the last years, numerous neural network frameworks have been developed to solve different tasks. Feedforward neural networks are the most straightforward manner to construct a neural network, where the information only flows unidirectionally from the input to the output neurons. Since images always come with a highly correlated pixel structure, it would be advantageous to consider this for the detection task. Fully connected structures would ignore the property by treating every pixel individually and also contain a large number of parameters. CNNs are able to consider these local pixel structures by using local receptive fields represented by filter masks with shared parameters, which can also come with spatial subsampling to reduce the spatial dimension. By creating a deep CNN with multiple layers, the network is able to learn the representation of features that are distinctive for a special property. This advantage over conventional computer vision approaches makes CNN very powerful in classification tasks. By extending the standard CNN architectures for classification by a localization part, the object detection task can also be solved. One popular architecture is the Single Shot MultiBox Detector (SSD) [1], which serves as the basic architecture of this work.

3 Network Architecture

Compared with other popular architectures, e.g., Faster R-CNN [6] or YOLO [7], the SSD architecture [1] distinguishes itself by achieving a proper trade-off between detection accuracy and computational speed. Generally speaking, the network is able to predict offsets to a default set of predefined bounding boxes for the localization as well as a classification score of how likely an object of interest is present within each box. The SSD network uses convolutional blocks of the popular VGG16 [8] as a base network to extract features. Additional convolutional layers are put on top to perform the high-level feature extraction and more down sampling. The prediction branches, consisting of convolutional filters for localization and classification, are then applied on several feature maps within the architecture. Experimental observation has shown that it is hard to detect small objects on the images. Vehicles or pedestrians far away from the camera only occupy a small number of pixels. In order to improve the performance of small object detection, the framework was modified the original SSD framework [1] by making more use of the feature maps used for the detection. The first new architecture uses the output of the last convolutional block within the base network (conv5_3) as additional layer for the predictions. It is called the SSD$^+$ network. The second uses the same number of feature maps for predictions as the original SSD architecture [1]. However, in this work, two prediction branches on top of block 4 were used – the SSD^2 network. Both approaches focus on improving the detection on small objects by using more information from the feature maps and predicting on a finer scaled set of bounding boxes. They come with a barely noticeable computational overhead compared to the original SSD architectures. Figure 2 gives a schematic overview of the network architectures.

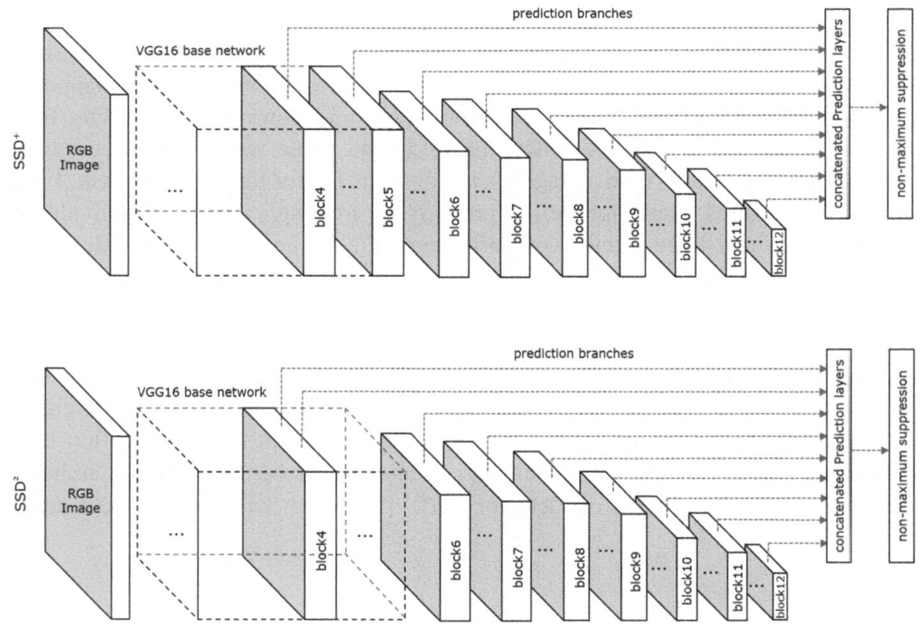

Fig. 2. SSD+ (top) and SSD^2 (bottom) architectures. The arrows represent the applied prediction branches on top of various feature maps.

4 Data Augmentation

Supervised learning is used to train the neural network to perform the detection and classification task. Therefore, a dataset with ground truth bounding boxes for each object that is to be detected needs to be available. Instead of creating our own image dataset for training and testing, we mainly used the public available Udacity [4] dataset. It consists of around 15000 images, having 146221 annotated objects for vehicles and pedestrians. In addition, the KITTI [2] and Caltech [3] dataset for object detections were examined in detail but were not used for training. Within the Udacity [4] dataset, 500 images were separated for testing and evaluation purposes and the rest of the dataset split in 80% for training and 20% images for validation. To increase the limited number of training examples, data augmentation is applied to synthetically generate more annotated data. The corresponding implementation uses a batch generator, which provides a batch of images to the network during training with an integrated data augmentation part. This makes it easy to store every dataset with its original images and perform the necessary data augmentation on the fly without creating distorted image and store them locally. The augmentation is done randomly and can consist of saturation, contrast and brightness variation, Gaussian blurring, scaling (smaller and larger), random size and fixed size cropping as well as flipping. Besides, the number of vehicles is far larger than that of pedestrians (see Fig. 3), which easily makes the training biased to vehicles. This leads to an even more challenging case for the pedestrian detection, which is in general very difficult due to the small size and gesture variations. To tackle this problem, a weighting scheme is applied during the matching process in training from ground truth and default boxes. It corrects the bias by rising the one hot encoded class vector value of the underrepresented class by a certain amount. With this, training on the unbalanced dataset is possible.

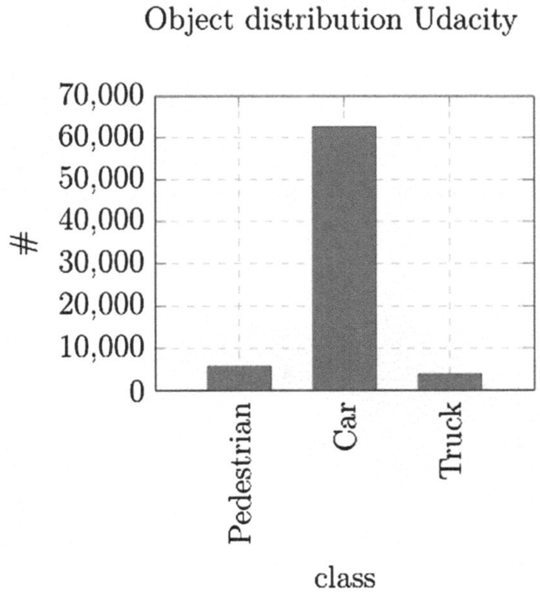

Fig. 3. Histogram for all objects appearing in the one of the Udacity [4] datasets

5 Bounding Box Generation

Since the SSD architecture makes predictions based on predefined set of bounding boxes, the generation of these so called *priors* is important. In order to understand the dataset better, a histogram of the sizes and the aspect ratios of the objects from the relative box height and width from the labels is first plotted. The results can be found in Figs. 4 and 5. The results show a bias towards small objects which makes the detection very difficult. As a result, the generation of bounding boxes for the detection not only needs to focus on smaller sizes but also has to cover large shapes for cars or trucks. Due to the extended SSD+ and SSD^2 architectures, a fine bounding box scale resolution could be used. The experiments conducted by the authors used boxes with relative scales of $\{0.05, \ldots, 0.90\}$. The vehicles appear in a wider rectangular shape whereas the pedestrians are taller in their shape. This results from the aspect ratio histogram shows that the main ratios of the width versus height for each bounding box are $\{1:5; 1:4; 1:3; 1:2; 1:1, 2:1\}$. With this information, the bounding boxes could be tailored according to the specific object shapes in the context of autonomous driving.

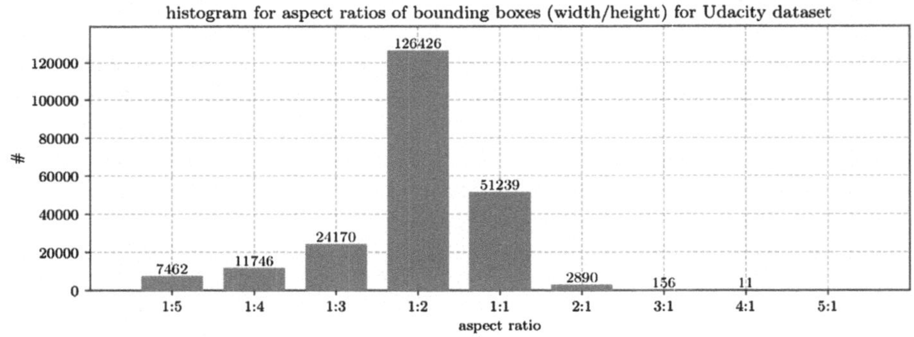

Fig. 4. Histogram chart shows the appearing aspect ratios (width/height) for the Udacity [4] dataset

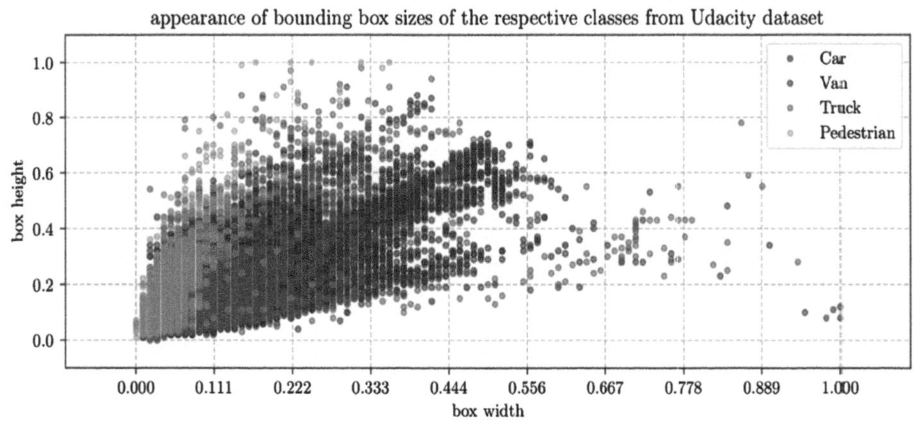

Fig. 5. Scatter plot for the distribution of bounding box sizes from all classes within the Udacity [4] datasets

6 Results

To train the neural network, transfer learning was used by adopting the weights of a pre-trained SSD architecture by Balanca [9]. The parameters of other layers of the proposed NN are randomly initialized. During the training, the batch size was set to 8, the epoch number to 20 and Stochastic Gradient Descent (SGD) was used as optimizer. 3000 iterations were executed during each epoch. This results in passing 24000 images per epoch and 480000 during the whole training through the network. After one complete iteration through the training image set, it is shuffled.

Fig. 6. Visualization of IoU calculation. Left: ground truth bounding box (green) and predicted bounding box (red). Right: calculation of the IoU value based on the bounding box areas. (Color figure online)

The performance of different network architectures and training configurations are given in Table 1. For testing, 500 images were taken from the Udacity [4] dataset. As performance metrics, an average and a mean average precision score, referenced as AP and mAP, respectively, were used. By using the intersection over union (IoU) criteria for the predicted and the ground truth bounding box (see Fig. 6), it is possible to distinguish between a true positive (tp) and false positive (fp) detection. With these parameters, a precision value can be calculated according to Eq. (1). A valid detection (tp) is present, if IoU > 50%.

$$precision = \frac{tp}{tp + fp} = \frac{\#\ of\ correctly\ detected\ objects}{\#\ of\ all\ objects\ in\ the\ testing\ set} \tag{1}$$

By averaging the precision values over all classes, one gets the mAP according to Eq. (2)

$$mAP = \frac{1}{N} \sum\nolimits_{i=0}^{N} precision_i \tag{2}$$

To have a baseline for the performance comparison, the original input sizes of [1] were used. Table 1 shows that the original models SSD300 and SSD512, with an input image size of 300 × 300 and 512 × 512 respectively, proposed from [1] show relative poor performance on the mAP score. The SSD300 in general runs at a very high framerate of 40FPS but only scores around 23% mAP. The extended SSD300[2] and SSD300$^+$ architectures improved the results but the performance was still not sufficient. The standard SSD512 architecture could already achieve a mAP score of more than 56% with a framerate of 22FPS. The proposed SSD512^2 and SSD512$^+$ networks can improve the detection performance of the detection by around 14% to a mAP score of 70%. Additionally, the SSD512$^+$ achieves the same FPS as the original model from [1]. By fine tuning the training configuration in terms of default bounding box generation, the SSD512$^+$ can improve the mAP by another 3% (see model 8 in Table 1). Since the SSD300 and SSD512 models form [1] take a squared input image but the datasets provide non quadratic images, the objects get distorted (squeezed) by reshaping the original image to the model input dimension. This is especially problematic for the pedestrians, since they have a very slim appearance. Due to the fully convolutional architecture of the SSD networks, the input size is flexible. Model 9 has a specific input size in order to not have the disadvantage of distorted objects coming from the resizing process. The SSDudacity uses a scaled input shape of a WUXGA image (1920×1200 pixels) to work best on the Udacity [4] images. This leads to no object distortions introduced by the downscaling. The result achieves a mAP score of 78.19% and the average precision score of the pedestrian detection increased by 10% compared to the best SSD512$^+$ architecture. The networks were trained and evaluated on a Nvidia GTX1080Ti GPU system using the Tensorflow™ framework from Google.

Table 1. Performance of various SSD architectures for vehicle and pedestrian detection

#	Model	Testing set	Average precision		mAP (%)	FPS
			'Vehicle' (%)	'Pedestrian' (%)		
(1)	SSD300	Udacity	38.55	8.49	23.52	40
(2)	SSD300^2	Udacity	40.69	8.39	24.61	39
(3)	SSD300$^+$	Udacity	40.22	12.78	26.50	39
(5)	SSD512	Udacity	77.83	34.83	56.33	22
(6)	SSD512^2	Udacity	81.58	58.77	70.17	21
(7)	SSD512$^+$	Udacity	81.21	58.87	70.03	22
(8)	**SSD512+**	**Udacity**	**82.81**	**63.42**	**73.19**	**22**
(9)	**SSDudacity**	**Udacity**	**82.90**	**73.46**	**78.18**	**22**

7 Conclusion

It is challenging to develop a single NN to perform object detection of more than one class. To achieve a high detection accuracy, traffic datasets with 'vehicle' and 'pedestrian' object annotations like in the Udacity [4] dataset were investigated to obtain an overview of the object numbers of each class, as well as the size and aspect ratios of the objects. The object sizes are of particular importance for the NN architecture, since the SSD network predicts offsets to a predefined set of bounding boxes. Since the datasets used were recorded in real world traffic environments, there is a huge bias towards vehicle objects in the images. In order to train a single network to detect vehicles and pedestrians, the training is modified to learn equally on both classes. This is done by data augmentation and a weighting scheme during the matching process from ground truth and default boxes. This combination enables a unified training pipeline for the imbalanced classes. One big challenge in the object detection task for autonomous driving is the variety of object sizes with a bias towards small objects. By using additional layers and a larger set of default bounding boxes, the architecture becomes more capable of detecting smaller vehicles and pedestrians with higher confidence scores. The performance of the proposal network was tested, and much better performance is demonstrated in comparison with the original SSD300 and SSD512 architectures [1].

8 Detection Examples

See Fig. 7.

Fig. 7. Detection examples of the SSD$^+$ architecture on various testing images [3, 10]. Vehicle detection are visualized as yellow boxes, pedestrian detections as purple boxes (Color figure online).

References

1. Liu, W., et al.: SSD Single Shot MultiBox Detector (2015)
2. Geiger, A., Lenz, P., Urtasun, R.: Are we ready for autonomous driving? The KITTI vision benchmark suite. In: IEEE Conference on Computer Vision and Pattern Recognition (CVPR) (2012)
3. Dollar, P., Wojek, C., Schiele, B., Perona, P.: Pedestrian detection: an evaluation of the state of the art. IEEE Trans. Pattern Anal. Mach. Intell. **34**(4), 743–761 (2012)
4. Cameron, O., Higgins, M., Gonzalez, E.: We're building an open source self-driving car (2016). https://github.com/udacity/self-driving-car. Accessed 09 Sept 2017
5. Trimble, T.E., Bishop, R., Morgan, J.F., Blanco, M.: Human Factors Evaluation of Level 2 And Level 3 Automated Driving Concepts: Past Research, State of Automation Technology, and Emerging System Concepts (2014)
6. Ren, S., He, K., Girshick, R., Sun, J.. Faster R-CNN: towards real-time object detection with region proposal networks (2016)
7. Redmon, J., Farhadi, A.: YOLO9000: Better, Faster, Stronger (2016)
8. Simonyan, K., Zisserman, A.: Very deep convolutional networks for large-scale image recognition. arXiv preprint arXiv:1409.1556 (2014)
9. Paul, B.: Repositories. https://github.com/balancap. Accessed 21 Sept 2017
10. Wojek, C., Stefan, W., Bernt, S.. Multi-cue onboard pedestrian detection. In: IEEE Conference on Computer Vision and Pattern Recognition (2009)

Tagungsbericht

4. ATZ-Fachtagung Fahrerassistenzsysteme – Von der Assistenz zum automatisierten Fahren

Mathias Heerwagen(✉)

Springer Fachmedien Wiesbaden GmbH, Abraham-Lincoln-Straße, 46, 65189 Wiesbaden, Germany
mathias.heerwagen@springer.com

Zusammenfassung. 220 Teilnehmer, und damit deutlich mehr als im Vorjahr, kamen am 18. und 19. April 2018 im gerade erst eröffneten Rhein-Main Congress Centrum in Wiesbaden zusammen, um sich über Neuigkeiten rund um die Themen Assistenzsysteme, automatisiertes Fahren und autonome Fahrzeuge zu informieren und auszutauschen. 22 Aussteller begleiteten die Tagung mit ihren Ständen und Exponaten.

1 Umwälzungen

Nichts weniger als „die nochmalige Erfindung des Automobils“ beschrieb Michael Hafner von Daimler in seiner Keynote, die mögliche Wege hin zum autonomen und unfallfreien Fahren aufzeigte. Der Vortragstitel sei nicht zu hoch gegriffen, wenn man an die Umwälzungen bei Sensoren, Aktoren und Rechenleistung denke, so Hafner. Um dem Endkunden sichere Systeme anbieten zu können, sind jedoch zuvor umfangreiche Tests notwendig. Daimler habe etwa zehn Millionen Testkilometer in Europa, Australien, den USA und China absolviert, um auch Szenarien zu entdecken, „die einem am Schreibtisch nicht einfallen“. In Melbourne beispielsweise gebe es für Europäer ein ungewöhnliches Verkehrsszenario, konstatierte Hafner, das man normalerweise nicht bedenkt: „Man muss links ausscheren, um rechts abzubiegen“. Das macht die Prüfprozeduren und Testfälle so kompliziert, bevor VDA-Stufe-3- und -4-Systeme auf den Markt kommen könnten. Abschließend stellte Hafner fest: „Wir müssen Vertrauen in Technik schaffen“, wenn es um das autonome Fahren geht. Dann biete sich dem Kunden eine völlig neue Mobilitätserfahrung im dritten Raum: Neben dem Zuhause und dem Arbeitsplatz werde auch das Autoinnere zum Erlebnis. Ralph Lauxmann von Continental entwarf in der zweiten Keynote eine neue Erlebniswelt in der Großstadt. „Dazu entwerfen wir Mobilitätskonzepte mit einem multimodalen Ansatz“, führte Lauxmann weiter aus. Mit autonomen People Movern und elektrischen Robotaxis könne es gelingen, die Unfallzahlen zu reduzieren, den

T. Bertram (Hrsg.): *Fahrerassistenzsysteme 2018*, Proceedings, S. 243–245, 2019.
https://doi.org/10.1007/978-3-658-23751-6_21

Stress beim Fahren zu senken und die Luftverschmutzung in urbanen Räumen zu verringern. Diese beiden neuen Fahrzeugklassen seien zwischen dem individuellen Verkehr mit dem Privatauto und den öffentlichen Bussen und Bahnen das geeignete Bindeglied, um den Verkehr besser und nahtlos zu gestalten.

2 Spannungsfeld aus politik, recht und ethik

In der Podiumsdiskussion am Ende des ersten Tags diskutierten Helge Kiebach vom KTI, Dr. Stephan Hönle von Bosch, Dr. Thorsten Leonhardt von Audi sowie Prof. Andre Seeck von der Bundesanstalt für Straßenwesen (BASt) die technische Entwicklung des automatisierten

Fahrens im Spannungsfeld von Politik, Recht und Ethik. „Solange es Mischverkehr gibt, wird es Unfälle geben", sagte Andre Seeck während der Diskussion. Dieser Einschätzung schließen sich nicht nur Ingenieure, sondern auch Versicherungsspezialisten an. Im mit regem Interesse verfolgten Versicherungsstrang erklärten Experten von der R+V-Versicherung und der Huk Coburg unter anderem, dass mit zunehmender Automatisierung die Unfallzahlen zwar sinken werden. Aufgrund der komplexen Technik und teurer Sensoren würden die Schadenaufwendungen jedoch gleich bleiben oder sogar steigen. Diskutiert wurden zudem Fragen zur Herstellerhaftung, wenn durch Systemversagen ein Unfall verursacht wird. Auch die Transport- und Logistikbranche wird sich in Zukunft intensiver mit rechtlichen Themen befassen müssen, sollte etwa das derzeit getestete Lkw-Platooning eines Tages zum Alltag gehören. In einer Sache waren sich die Versicherungsexperten einig: Die vielfach erwähnte Vision Zero sei kaum zu erreichen. Damit man diesem Ziel aber möglichst nahe kommt, braucht es präzise Sensoren. Um deren effiziente Entwicklung mithilfe von Rohdatenschnittstellen, die Verbesserung des Umgebungsmodells durch die Vergrößerung der Sensorreichweite und um die Anforderungen zukünftiger Radarsysteme ging es beispielsweise in der Sensor-Session. Im Bereich Safety & Security erläuterten Referenten von Etas, Escrypt und Continental unter anderem, wie sich eine Beobachterstruktur herleiten und anwenden lässt, um Inkonsistenzen innerhalb einer statischen Umfeld-modellierung zu erkennen. In der Session Neue Methoden referierten Experten über Methoden zur cloud-basierten Validierung und Optimierung von hoch automatisierten Fahrzeugen, über systemisch generierte Tests für komplexe Fahrszenarien oder Connected Development für Fahrerassistenzsysteme.

3 Autonomes fahren im blick

Auf der 4. Internationalen ATZ-Tagung „Fahrerassistenzsysteme – Von der Assistenz zum automatisierten Fahren" zeigte sich, dass viele Assistenzsysteme mit Sensorik und Aktorik soweit up to date sind, dass sie als serienreif bezeichnet werden können. Das nächste Ziel, autonomes Fahren, haben alle Autohersteller und deren Zulieferer im Blick. Im kommenden Jahr spiegelt sich dieser Trend daher auch in einem neuen

Tagungsnamen wider. 2019 lädt das ATZlive-Team für 2. und 3. April 2019 wieder ins Rhein-Main- Gebiet ein, dann zur 5. Internationalen ATZ-Tagung mit dem neuen Motto: „Automatisiertes Fahren – Von der Fahrerassistenz zum autonomen Fahren“.

Quelle: Heerwagen, M. ATZ Automobiltech Z (2018) 120: 80. https://doi.org/10.1007/s35148-018-0086-4.

Autorenverzeichnis

T. Bertram (Hrsg.): *Fahrerassistenzsysteme 2018,* Proceedings, S. 247–248, 2019.
https://doi.org/10.1007/978-3-658-23751-6

The manufacturer's authorised representative in the EU is Springer Nature Customer Service Centre GmbH, Europaplatz 3, 69115 Heidelberg, Germany. If you have any concerns regarding our products, please contact ProductSafety@springernature.com

Printed and bound by CPI Group (UK) Ltd, Croydon, CR0 4YY
19/08/2026
02200191-0009